D0010763

TEACHING AND LEARNING SCIENCE
A Handbook

VOLUME 2

Edited by Kenneth Tobin

PRAEGER

Westport, Connecticut
London

CATALOGUED

Library of Congress Cataloging-in-Publication Data

Teaching and learning science : a handbook / edited by Kenneth Tobin.
    p. cm.
  Includes bibliographical references and index.
  ISBN 0–313–33573–7 (set : alk. paper)—ISBN 0–313–33936–8 (v. 1 : alk. paper)—
ISBN 0–313–33937–6 (v. 2 : alk. paper) 1. Science—Study and teaching.
1. Tobin, Kenneth George, 1944–
  Q181.T3513 2006
  507.1'2—dc22   2006015088

British Library Cataloguing in Publication Data is available.

Library of Congress Catalog Card Number: 2006015088
ISBN: 0–313–33573–7 (set)
     0–313–33936–8 (vol. 1)
     0–313–33937–6 (vol. 2)

First published in 2006

Praeger Publishers, 88 Post Road West, Westport, CT 06881
An imprint of Greenwood Publishing Group, Inc.
www.praeger.com

Printed in the United States of America

The paper used in this book complies with the
Permanent Paper Standard issued by the National
Information Standards Organization (Z39.48–1984).

10 9 8 7 6 5 4 3 2 1

Contents

PART 6

MAKING CONNECTIONS WITH SCIENCE

33

Culturally Relevant Pedagogy for Science Education

Deborah J. Tippins and Scott Ritchie

It is not surprising that students frequently report that what they learn in school has little relevance to their lives outside the classroom or to their futures (Aikenhead 1996). This is particularly the case in school science, which typically reflects middle-class experiences and excludes the lives of students on the margins (Fusco 2001). There is a growing concern among science educators regarding how increasingly diverse student populations connect their lives to science. As science educators have attempted to confront the complex issues surrounding notions of multicultural science education, they have struggled with creating a "culturally relevant pedagogy" for science teaching and learning. Science education research over the past fifteen years points to many attempts to infuse cultural relevancy and identify pedagogies that foster the academic success of marginalized students in science. However, much of the research in science education has focused on psychological analyses of learning, specifically attending to issues of conceptual understanding by individuals or groups of students. More recently, with the call for more sociocultural analyses of science learning, there have been attempts to understand "relevancy" by venturing into students' and teachers' lifeworlds beyond science classrooms, including places such as lunchrooms or homeless shelters. Gale Seiler's work (2001) with African American youth in urban Philadelphia and Angela Calabrese Barton's studies (1998) with homeless families reflect such efforts to move science education research beyond the four walls of the classroom.

CULTURALLY RELEVANT PEDAGOGY:
TRADITIONAL PERSPECTIVES

The term *culturally relevant* originally stems from Gloria Ladson-Billings's (1995) accounts of culturally relevant pedagogy practiced by exemplary teachers of African American students. Ladson-Billings characterized culturally relevant teachers as ones who use students' home culture as a basis for helping them examine and critique social inequality and work for social change. By extension, she described culturally relevant teaching as a practice that calls on teachers to create democratic and multicultural classrooms that empower students by drawing upon their culture. Numerous studies in science education have extended the work of Ladson-Billings across a variety of contexts and communities, with a renewed vigor.

RECENT PERSPECTIVES ON CULTURALLY
RELEVANT SCIENCE TEACHING AND LEARNING

What do culturally relevant science teaching and learning look like in practice? Angela Calabrese Barton (2000) suggests that the many attempts to infuse culturally relevant pedagogy in pre-K–12 classrooms and science teacher preparation programs meet with failure because "science education as a community does not have a clear idea of what multicultural science teaching and learning looks like in practice" (800). Some of the various discourses used to promote "culturally relevant," "culturally congruent," or "culturally responsive" pedagogies include the following.

Crossing Cultural Borders

The teaching and learning of school science represent a particular culture of science that may not fit with the everyday lifeworld of students and teachers. Glen Aikenhead (1996) points out that in order for students to experience success in the culture of school science, they may need to negotiate "cultural borders" such as language conventions and epistemologies. In his work with First Nations people in Canada, Aikenhead found that border crossing into the culture of school science was not easy for many students. Building on the work of Albert Costa, he uses a framework to analyze how students move from a lifeworld outside of school into the culture of school science. He describes students who found both school and science relevant to their family culture as *smooth border crossers* and "Potential Scientists." At the other end of the spectrum, some students' lifeworlds were strongly at odds with the values and beliefs implicit in both school and Western science. School science may hold little relevance in the lives of these "Inside Outsider" students, and some may even experience overt forms of discrimination. As a discourse for thinking about relevance, the idea of "cultural borders" suggests that the epistemological basis of science, itself, remains unchallenged.

Drawing on Funds of Knowledge

Pauline Chinn's (2003) work with Asian Hawaiians draws on "community funds of knowledge" as a source for creating a culturally relevant science curriculum. Chinn emphasizes that cultural dialogue and knowledge of students' lifeworlds are important precursors to using Native Hawaiian knowledge as a centerpiece for creating a relevant curriculum. Similarly, Lorrie Hammond's (2001) school-community garden project and field house were designed to empower Iu Mienh immigrant families (from Southeast Asian hill tribes displaced from Laos as an aftermath of the Vietnam War) with respect to science education by bringing their "funds of knowledge" into the science curriculum. Hammond and her university students collaborated with Iu Mienh refugee families to develop a Southeast Asian garden that connected their science learning with aspects of their cultural capital—increasing the likelihood that they would learn science in ways that were perceived as relevant and understandable.

Creating a Practicing Culture of Science Learning

Based on her work with urban teenagers in a community gardening project, Dana Fusco (2001) suggests that relevancy is inherent in science learning when students become producers of science. The importance of creating a "practicing culture of science learning" is at the heart of her way of thinking about culturally relevant pedagogy. Fusco's notion of a practicing culture is one in which "children draw on as well as define science, its activities and its uses within a particular context for specific purposes" (862). In relation to various discourses of educational relevance, Fusco argues that what is learned is not as important as how it is learned, and emphasizes that in a practicing culture of science learning, "science can be learned even when one's experiences in science are at odds with one's experiences outside science" (862).

Thinking Critically about Knowledge and the World

Many science educators argue that education, rather than merely reproducing status quo social inequities, should strive toward social change. From this perspective, not only should teachers make the standard curriculum more culturally relevant for their students; but they should also foster a critical stance toward the curriculum—its origin, whose ideas it represents, who it leaves out, and so on. In this social justice discourse of relevancy, active problem posing by students replaces the passive reception of a "banking" or "transmission" approach to science in which the ever-knowing teacher deposits information into her students' heads, and the science curriculum is cogenerated by students and teachers alike. Barton and Osborne's work in after-school science programs and homeless shelters (2001) calls for the building of science centered around children's informed knowledge and experience. Barton emphasizes that children and teachers, as coproducers of science, should explore multiple methods of talking, thinking, and doing science—ways of knowing that may stand in stark contrast to what is often perceived to be the nature of "real" science.

MOVING BEYOND CULTURALLY RELEVANT PEDAGOGY

As science educators have attempted to make connections between the science curriculum and the lives of their students, they have begun to pose the question "Relevant science education, but relevant to what?" We consider this question and attempt to move the multicultural science education dialogue of recent years beyond traditional notions of "culturally relevant," "culturally responsive," or "culturally congruent" science teaching and learning through two examples.

RELEVANCY AS CURRICULUM-CENTERED SCIENCE

As a teacher of predominantly Latino students, Ms. Jennings wants to make her science curriculum more culturally relevant to the lives of her students. She collaborates with her school's English as a Second Language (ESL) teacher to gather information and solicit input from a variety of education stakeholders, including the state curriculum, her students and their parents, other community members, teachers, professional development literature, and research on effective instruction for nonnative English speakers. Using the information she gathers, she starts with the prior learning experiences and interests of her students to develop a science unit that integrates many literacy process activities and skills while staying focused on teaching about the earth's history, a state science objective for her grade level. Students learn scientific processes and concepts, and strengthen their command of English as well—a goal identified by parents and community members.

Commentary

This approach begins with a teacher who, after seeking input from various education and community sources, abstracts the science curriculum from the information she gathers. The manner in which the curriculum is presented and taught is adapted to meet local needs, but the scientific epistemology remains unchanged. Now let us examine an approach that starts to make a transition beyond culturally relevant curriculum.

RELEVANCY AS COMMUNITY-BASED SCIENCE

Like Ms. Jennings, Ms. Harper is interested in making her science curriculum more relevant to her students' community. She too gathers information from others before implementing her curriculum. At the beginning of the year, Ms. Harper meets with parents to learn more about each child and the parents' hopes for their child. Many of these meetings occur at the child's home. While it is not required, Ms. Harper also stays involved in her students' lives and communities by attending PTA meetings, activities at the neighborhood community center, and extracurricular sporting events. During visits to her students' homes, Ms. Harper notices that inadequate heating is an issue for her students' families. In many of the homes she visits, multiple family members sleep in the living room to stay warm. Kerosene heaters are common. Some parents offer Ms. Harper blankets during their meetings, apologizing for the cold conditions. Parents discuss how poor insulation in their factory housing—common for students at Ms. Harper's school—creates a cold, drafty living environment. Ms. Harper decides to invite her students and their families to investigate why their homes are cold. She holds a meeting with students and parents to discuss the issue, and together they co-develop an inquiry-based curriculum that explores how their homes are insulated, what other insulation options exist, and the effectiveness of each type. Drawing upon community resources, Ms. Harper enlists a home improvement store employee to bring in various insulation materials so her students can conduct experiments to see which materials best prevent heat loss. This semester-long thematic unit culminates in a community effort to reinsulate several of the children's homes.

Commentary

In this approach to science, the teacher co-constructs the curriculum with members of the community itself rather than abstracting the curriculum from stakeholder input. What emerges is immediately relevant to students, parents, and other community members. However, in this approach the teacher is still the principal instigator of the curriculum. Without the teacher, the community would not have a part in constructing the curriculum. Such an approach maintains an epistemological structure of a teacher-researcher investigating a phenomenon, as well as dependent relations between researcher and researched or teacher and community. It is our hope that other models will be developed that move beyond these shortcomings.

RETHINKING CULTURALLY RELEVANT PEDAGOGY

We maintain that additional approaches are needed to make science truly relevant to students. While we may tinker around with the existing curriculum or even invite students and members of their communities to codevelop science curricula and pedagogy, such approaches are limited at best. This is not to say that such approaches are not necessary; indeed, they move us beyond what others call a "banking" approach toward instruction. However, for science to be relevant to its practitioners, *origin* is of great importance. Where does science originate? Whose interests does it serve? To what extent are those who practice science able to originate future scientific investigations without depending on others to frame the experience? Questions such as these have the potential to move us beyond mere cultural relevance to science education that both emerges from and reports to the community it is designed to serve.

REFERENCES

Aikenhead, Glen. 1996. Science Education: Border Crossing into the Subculture of Science. *Studies in Science Education* 27:1–52.

Barton, Angela C. 1998. Teaching Science with Homeless Children: Pedagogy, Representation and Identity. *Journal of Research in Science Teaching* 35:379–94.

———. 2000. Crafting Multicultural Science Education With Preservice Teachers Through Service-Learning. *Journal of Curriculum Studies* 32:797–820.

Barton, Angela C., and Margaret D. Osborne, eds. 2001. *Teaching Science in Diverse Settings: Marginalized Discourses and Classroom Practice.* New York: Peter Lang.

Chinn, Pauline. 2003. A Hawaiian Sense of Place: Science Curricula Incorporating Hawaiian Ways of Knowing. Paper presented at the National Association for Research in Science Teaching Conference, Dallas, TX, April.

Fusco, Dana. 2001. Creating Relevant Science through Urban Planning and Gardening. *Journal of Research in Science Teaching* 38:860–77.

Hammond, Lorrie. 2001. Notes from California: An Anthropological Approach to Urban Science Education for Language Minority Families. *Journal of Research in Science Teaching* 38:983–99.

Ladson-Billings, Gloria. 1995. Toward a Theory of Culturally Relevant Pedagogy. *American Educational Research Journal* 32:465–91.

Seiler, Gale. 2001. Reversing the Standard Direction: Science Emerging from the Lives of African American Students. *Journal of Research in Science Teaching* 38:1000–14.

Interactive Historical Vignette: The Scientific Endeavors of Mary Anning, the First Woman Paleontologist

James H. Wandersee and Renee M. Clary

When asked to name a prominent scientist, most students will not identify a successful female. In the past, women were not even allowed to participate in scientific societies. It was not until 1945 that the first female was allowed entrance into the prestigious Royal Society of London. It is an unfortunate consequence of nineteenth-century society that *recorded* science was accomplished through socially accepted white males. This should not imply that women and working-class men were not conducting their own scientific investigations; indeed, William Smith, the marginalized creator of the first geological map of England, lamented in 1816 that the theory of geology was in the possession of one class of men, while the actual practice of geology was in the possession of another (Woodward 1907).

Working-class females faced two barriers: they were not only of the wrong social class to theorize about scientific practice, but they were also of the wrong gender. Yet, knowledge of some remarkable scientific females of this period survives. Mary Anning of Lyme Regis, England, is appropriately receiving some belated recognition.

MARY ANNING AS NINETEENTH-CENTURY PALEONTOLOGIST

Most of us have heard of Mary Anning, although we may not have been aware of her significance. Paul McCartney (1977) claimed Anning was the subject of the children's tongue twister, "She sells seashells on the seashore." The "seashells" immortalized in the verse are actually fossilized specimens of extinct animals, not the shells of recently living mollusks. Although Horace

Woodward (1907) called Anning "the most notable collector during the early part of the nineteenth century" (115), her peers did not consider Anning, as an "uneducated" woman, a prominent paleontologist.[1]

As a young girl, Anning collected fossils, or "curiosities," with her father from the cliffs near Lyme Regis, England. As the classic Anning story relates, upon her father's death, Mary began an entrepreneurial collecting endeavor in order to bring money into her mother's household. Anning's discoveries included the remains of an ichthyosaur (an extinct, dolphin-shaped marine reptile), a plesiosaur (an extinct long-necked marine reptile), and a pterodactyl (an extinct flying reptile). The fossils she found were sold to tourists who traveled to Lyme Regis, as well as to prestigious scientists and museums of the time. Simon Winchester (2001) noted that the recorded list of Anning's customers included practically all of the prominent geologists of the period.

EARLY LIFE

Anning was born to Richard and Mary Anning of Lyme Regis in 1799, and was the third "Mary Anning" to be named: the first Mary Anning was her mother, while the second Mary Anning was an elder sister who died in a house fire the year before Mary was born (Torrens 1995). Her well-used name was not the only unusual aspect of her early life: when only a year old, Mary was with her nurse and two other children when a downpour forced the group to find refuge under a tree. Lightning struck the tree, killing all but Mary. She was pulled from beneath her dead nurse's body, and revived in warm water. Many years later, when Roberts (1834) wrote an early history of Lyme Regis, he attributed the lightning strike as the source for Anning's intelligence; whereas she had been quite dull before the incident, afterwards she exhibited an intelligence lasting for the rest of her life. It is perhaps not unusual that nineteenth-century society insisted on a supernatural source of Mary Anning's intelligence, given that women were not perceived as being the possible intellectual equals of men.

FOSSIL DISCOVERIES AND FAME

When Mary Anning was only ten years old, her father's untimely death greatly affected the family's resources. Mary, her brother, and her mother continued the fossil recovery business as a source of income, with Mary as the primary fossil collector. However, early fossil successes were rare, and the family was often in need of parish relief.

Most of the fossils recovered by Mary Anning were ammonites and belemnites (extinct cephalopods similar to the chambered nautilus and squid of today), and other types of mollusks. However, her extinct reptile fossils were to secure Mary Anning's reputation among scientists. Although the recovery of the first ichthyosaurus in 1811 is often attributed to Anning, her brother, Joseph, probably made the first discovery of the skull (see Figure 34.1).

FIGURE 34.1 Reproduction of *A School-girl Meets Ichthyosaurus* by Charles Edmund Brock, printed in Mee's (1925) *Children's Encyclopedia* and reproduced by Torrens (1995) in the *British Journal for the History of Science*. The illustration incorrectly portrays Mary Anning as the finder of the ichthyosaurus, although actually it was her brother Joseph who discovered the skull. The specimen also was not discovered intact, contrary to this illustration, and we now know that Mary Anning found the rest of this fossil a year later.

However, Mary, intent to find the remainder of the skeleton, diligently searched the cliffs for an entire year before she met with success. The location of the skeleton—high in the cliffs—forced young Mary to use her fossil proceeds to hire local men to recover it from the cliff (Goodhue 2004). A decade later, Anning made her second major reptile fossil find: the plesiosaur, a long-necked marine reptile completely new to science. Torrens (1995) reported that, in the eyes of her scientific colleagues, this discovery was her greatest.

FIGURE 34.2 Ichthyosaurus specimens located in the Hall of Marine Reptiles at the Natural History Museum of London, which has several of Mary Anning's fossil discoveries on display. Anning is now identified by the signage as the collector of some of her specimens. Only the owners had been credited in the past.
Photograph by R. M. Clary.

Her third important fossil discovery was the first British example of the flying reptile known as a pterosaur. Anning recovered several beautifully preserved fossilized reptiles; her patience and perseverance at freeing the entombed animals from their matrix are visible in the fine detail of the fossils. Many of Mary Anning's fossils can be seen today at the Natural History Museum of London's Hall of Marine Reptiles (see Figure 34.2).

THE INTERACTIVE HISTORICAL VIGNETTE

Since contributions by women in sciences have largely been ignored, we believe that the introduction of Mary Anning as a female role model is an appropriate addition to today's science classroom. Anning's struggles as a working-class female in a male-dominated scientific community illustrate the need for an inclusive perspective of the nature of science. We think it is important for students to understand not only what scientists know, but also how science works. Science was not always as open to the ideas and contributions of women, minorities, and the working class as it is today.

The Interactive Historical Vignette (IHV) was developed by Wandersee in 1989 to facilitate the inclusion of the nature of science in the classroom.

FIGURE 34.3   Images of a young boy and girl that can be utilized for Mary and Joseph Anning masks. The boy's image was drawn in 1904 by artist David Foggie (1878–1948). Both head images are to be enlarged by photocopier to about human size, cut out, and mounted, via glue stick, to a 3′ + 3′ white foam board square. The masks are to be held in front of the actors to cover up their own heads. It also helps to rock the masks slightly from side to side during the characters' stage conversation, to make the conversation appear animated.

Because teachers must work within an already packed science reform curriculum, the IHV is designed to require minimal classroom time, capitalize upon students' interest in stories and drama, and engage the audience in predicting how the story will end. The scripts are written to be docudramatic in nature and represent possible conversations, based on histories of real scientific events. Ideally, inclusion of IHVs in the science classroom should not be a sporadic or one-shot treatment of the nature of science. IHVs should be incorporated weekly, in 10–15 minute blocks, with each vignette focusing on a different figure in the history of science.

In each IHV, a single facet of the nature of science—for example, skepticism, respect for the power of a theory, thirst for knowledge, or willingness to change one's opinion—is central to the lesson. The storyline of each vignette is based upon a pivotal incident drawn from the history of science, and focuses upon an intellectual or behavioral choice in the life of a scientist. For a complete description of the construction of an IHV and the twenty critical attributes of scientific thought, see Wandersee and Roach (1998).

IHVs can be implemented in various ways in the science classroom. Teachers can dress in period costume and use scientific props, or simply manufacture large IHV character masks (see Figure 34.3). These character masks are typically constructed using large (2' × 2') photocopies of the head and shoulders of the scientist in the vignette, with inconspicuous eye, nose, and mouth openings cut into the image. Many copy shops can make these quickly on a copy machine with a zoom, or enlarge, function. Once the teacher has presented a regular series of IHVs during the first month of the course, students can then be assigned to construct their own vignettes; and, after the teacher has approved their content, they can present them to the class. We have also found that it is helpful to students to show them brief videotapes of exemplary presentations done by previous students.

MARY'S ICHTHYOSAURUS: AN INTERACTIVE HISTORICAL VIGNETTE SCRIPT

The following IHV takes place on the beach near Lyme Regis, a small community located in the far southwest of England. Can you find Lyme Regis on the map of England? The beach of Lyme Regis is a rocky coast with treacherous cliffs. Whereas many people today like to live along the coast with a view of the sea, the Lyme Regis coast was not an ideal site to build a house. During storms, parts of the cliff often collapsed to the beach below.

As our story opens, a ten-year-old girl and her older brother are searching for something along the beach. The young girl is Mary Anning. Have you ever heard that name before? Even if you have not heard the name *Mary Anning*, you probably have heard something about her. Do you remember the tongue twister "She sells seashells on the seashore?" That is about Mary Anning! However, Mary Anning and her brother, Joseph, weren't collecting seashells. They were collecting fossils, the ancient remains of animals that no longer exist. Mary and Joseph learned how to find fossils from their father. This is a fun activity for them, but also a necessity: after the death of their father two years ago, the family is often without enough money to eat, and the sale of these fossils to the tourists helps them survive. The year is 1811.

Joseph: Mary, quick! Come and see! I've never seen a skull this big before! And, look! The eye is huge!

Mary: Joseph, what kind of fossil do you suppose it is?

Joseph: I don't know, but it's the best thing we've ever found, even if it's just the skull. If only we could find the rest of it!

Mary: I'll search every day for it, Joseph. I know I'll find it! Father said to never give up. If you look long enough, you will find what you are looking for.

Stop: Ask the students to predict how the story will end. Was this an important discovery for Mary and Joseph Anning? For science? What do you suppose the animal was?

One year later:

> *Mary (out of breath):* Joseph. . . . Joseph! I found it!
>
> *Joseph:* Found what?
>
> *Mary:* The rest of the sea dragon! I found the rest of the sea dragon high in the cliffs! There are lots of vertebrae in its back, and it is huge! No wonder its eyes were so big! I told you I would find it if I kept looking every day. It wasn't where we thought it would be, but I found it! *(She does a dance of joy.)*
>
> *Joseph:* But, Mary, how are we going to get it out of the rock if it is high in the cliffs?
>
> *Mary:* I thought about that, Joseph, when I was running home to tell you! We'll use some of the fossil money to hire workmen to get it out for us! Surely, we will be able to sell it to the lord of the manor or some other scientist for a large sum of money!
>
> *Joseph:* That's a great idea, little sister! Mother will be so excited!

> *Stop:* What characteristics of a scientist does Mary Anning exhibit? (A few examples are persistence, keen observation, and willingness to change her opinion as to where the rest of the fossil might be.) Do you think the fossil really is a "sea dragon"? Do you think animals like this one are still alive today? Why would it have such large eyes? Would you have searched for an entire year to locate the rest of a fossil? Why do you think Mary was so happy when she found it?

Although Mary Anning was only ten years old when the skull was found, she continued to search for fossils along the three miles of cliffs near Lyme Regis for the rest of her life. She found many important fossils of extinct animals, and she became famous for her discoveries and her knowledge of fossils. Some researchers believe that Mary Anning was the first woman paleontologist in the world.

ADDITIONAL ANNING RESOURCES

In an effort to celebrate the accomplishments of Mary Anning and to promote recognition of her geologic contributions, a special day is set aside each March by the Lyme Regis Museum and the Jurassic Coast World Heritage team as "Mary Anning Day." The renaissance of interest in Mary Anning's life in recent years also has resulted in several books written about her life and work. For younger readers, Don Brown's (1999) *Rare Treasure: Mary Anning and Her Remarkable Discoveries* allows children a glimpse into the life of a remarkable scientist of the nineteenth century. Brown's book is also the recipient of the 2002 Giverny Award, which is given annually to the best children's science picture book (see www.15degreelab.com). A recent addition to Mary Anning bibliographies is Goodhue's (2004) *Fossil Hunter: The Life and Times of Mary Anning (1799–1847)*, which can be obtained from www.amazon.co.uk. An additional Mary Anning IHV may be found in *American Biology Teacher* (Clary and Wandersee 2006).

Often our students are presented with "bare bones" science, and do not get a glimpse into the men's and women's lives that helped to influence the evolving discipline. We believe that the incorporation of the history of science, as well as the characteristics of a scientific mind, through Interactive Historical Vignettes can provide opportunities to actively engage students and illuminate the social, historical, and political contexts in which science developed.

NOTE

1. Horace Woodward chronicled the early history of the Geological Society of London, as well as the lives of some of the early prominent geologists. Woodward's texts, printed in the early 1900s, are an invaluable reference when researching the interactions between the prominent geologists of the day and the individuals with whom they interacted, such as the prominent fossil collector, Mary Anning.

REFERENCES

Brown, Don. 1999. *Rare Treasure: Mary Anning and Her Remarkable Discoveries.* Boston: Houghton Mifflin.

Clary, Renee M., and James H. Wandersee. 2006. Mary Anning: She's More than "Seller of Sea Shells at the Seashore." *American Biology Teacher* 68:153–57.

Goodhue, Thomas W. 2004. *Fossil Hunter: The Life and Times of Mary Anning (1799–1847).* Bethesda, MD: Academica Press.

McCartney, Paul J. 1977. *Henry de la Beche: Observations on an Observer.* Cardiff, Wales: Friends of the National Museum of Wales.

Roberts, George. 1834. *The History and Antiquities of the Borough of Lyme Regis and Charmouth.* 2nd ed. London: Samuel Bagster & William Pickering.

Torrens, Hugh. 1995. Mary Anning (1799–1847) of Lyme: 'The Greatest Fossilist the World Ever Knew.' *British Journal for the History of Science* 28:257–85.

Wandersee, James H., and Linda M. Roach. 2005. Interactive Historical Vignettes. In *Teaching Science for Understanding: A Human Constructivist View,* edited by J. Mintzes, James H. Wandersee, and Joseph D. Novak, 281–306. San Diego, CA: Academic Press.

Winchester, Simon. 2001. *The Map That Changed the World.* New York: HarperCollins.

Woodward, Horace B. 1907. *The History of the Geological Society of London.* London: Geological Society Burlington House.

WEBSITES

The Philpot Museum website provides a calendar of events and announces when Mary Anning Day is celebrated each year. Photographs of previous Mary Anning Days are archived: www.lymeregismuseum.co.uk

The Mary Anning website at the University of California, Berkeley, provides a brief overview of Mary Anning's life. Active links to some of the fossil specimens she collected are included: www.ucmp.berkeley.edu/history/anning.html

35

Science and Art

Margery D. Osborne and David J. Brady

As people teaching others (children, college students, and teachers) about science, we work very hard to communicate certain core experiences—ones that enable discovery and understanding in science. We worry about how to represent these to others, how to help our students experience them for themselves, and finally about what they actually involve and signify. By that, we mean engaging the creative processes that lie behind the production of the facts, theories, and procedures all of us think of when the word *science* is said to us. In order to do this, we suggest that a comparison between the disciplines, methodologies, and cultures of arts and sciences is helpful. Victor F. Weisskopf (1979), a professor of physics at MIT and former president of the American Academy of Arts and Sciences, points to the "complementarity" (Niels Bohr, atomic physicist, in Weisskopf 1979, 474) of science and art. He views the two as different avenues of human creativity, with each representing different aspects of reality and each adding to our understanding of natural phenomena. Both give us deeper insights into our environment and provide meaning and sense to human experience.

We think that the link between art and science can be used as a catalyst for students to compare and think about the two by using both to explore the same objects and ask what they can find out or see about something using science or art. The juxtaposition is the starting point for reflection on central (and difficult) questions such as "What is art?" and "What is science?" And, at an even more basic level, it provokes conversations about the nature of seeing: what can be seen (and not seen) through art rather than science, and vice versa? What can be seen through the use of different media, and how is this related to what we want to see in the first place (Osborne and Brady 2000)? Both art and science involve close observation; both depend upon choice in the determination of procedure and medium (how much of this is conscious and how much unconscious is another matter); both depend on insight and inspiration but also hard work; both require persistence and tenacity as well as self-discipline; and both are creative endeavors.

To do this, we have been developing a curriculum in which we teach science and science pedagogy in unconventional ways with children and preservice and practicing teachers. We have purposely blurred the edges between science and art—doing science through art and art through science. In this process, we engage ongoing conversations about the integration of the arts into subject matter teaching (Eisner 1998). We feel that at a fundamental level, both science and art are about "seeing," seeing new things and seeing in new ways. This intersection is the starting point of our teaching, and in combining the two in our instruction we create a space in which we use the combination of the two to create a potential space of creative critique. We do this for two reasons: to cause students to think hard about the activities they are engaged in and to enrich their abilities in a process common to both disciplines, the ability to see. When viewed as an exploration of seeing and a development of the ability to see, science and art become tools that build upon and enrich each other.

In this chapter, we describe some of the spaces we have created where students of science can explore the meanings of the seeing process in the context of developing a rich understanding of the nature of science through engaging in art and science simultaneously. As we watch our students do this, we come to recognize that there is an array of "ways of seeing" in science, each of which enables the discovery of different qualities in an object. This in turn causes us to reflect back on the nature of the process and our purposes in engaging in it. In particular, we describe how we intertwine science with visual art in our classes to enable seeing. We present here our attempts to construct a "taxonomy" of ways of seeing. These are as follows:

- Looking closely/discovering the nature of things (by close observation and by comparison)
- Exploring the process of seeing
- Manipulating the image itself and discovering things
- Experimenting with seeing/seeing engendering "thinking-about"

There is a movement in this taxonomy between focusing upon the content of images—seeing content—and becoming aware and analytical about the processes of seeing itself, and then becoming intrigued, once again, by the object of the looking.

LOOKING CLOSELY/DISCOVERING THE NATURE OF THINGS

In our first example, first-grade children explore plants by observing them and also by growing them from seeds and bulbs. In this teaching, we facilitated their close observation and comparison of the plants and growing seeds through observational drawings. The drawings shown in Figure 35.1 were done using fine-point drafting pens and tools such as a magnifying lens and microscopes. Children were asked to make comparative drawings between types of plants chosen because of their contrasting aspects. They were also asked to make compilations of drawings over time to track the changes in seeds as they germinated and grew.

Here, scientific observation, description, and explanation are interlinked in the thinking of the student. We have been intrigued, through the years that we have taught in the sciences, by the difficulties we have experienced teaching students to see analytically. A large part of the science teaching that we have done has involved creating possible experiences that students can have in which they look at things, new things and things that they have seen many times before, and see them in new ways. These children are struggling with this process. They find they can identify a plant as a plant because they have seen one many times before in a holistic sense. Breaking this whole into

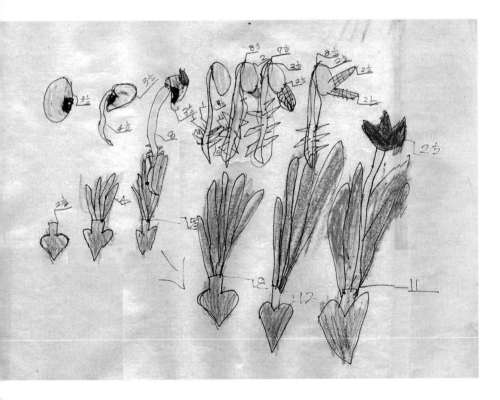

FIGURE 35.1

fragments that can be examined as components in isolation and as parts upon which the whole is contingent requires at least an implicit recognition by the individual of the mental and sensual processes of identification and naming. It is at this level that drawing becomes so important.

EXPLORING THE PROCESS OF
SEEING/PROCESS NOT CONTENT

In this example, first- and second-grade children analyze photographic representations of objects, looking for patterns and designs in the photograph. This process enables them to selectively see some qualities of the object while obscuring others. The children examine photographs, describing patterns and discovering how these patterns once articulated both enable and obscure seeing. In particular, we examined a collection of slides of close-ups of leaves, flowers, and bricks, things with repeating patterns. The children described the patterns, and they began to talk about how they could use the patterns to predict how things looked that weren't included in the picture. They also started to use mathematical relationships to describe pattern progressions. For example, Suni: "The bricks . . . that one has four holes and that one has three holes and that one has two holes, four minus one is three, minus one is two, I think the next one will have one or four and the pattern will start over." They extended their observations of patterns to talking about how the patterns might imply an explanatory function. Finally, we used patterns to make comparisons, examining a collection of dried leaves—maple, tulip tree, sweet gum, poplar, cut-leaf birch—that are similar yet different (Bateson 1979).

We had been looking at the picture shown in Figure 35.2. I have just asked the children what patterns they see. Shumshad says, "I know one, it was just a little one, it was, I know one that was [curved] . . . it goes like that and then like that." He makes gestures with his fingers in the air, indicating a series of large curved lines that repeat. I asked if others could see this. Many answered yes—and started listing other patterns as well: the repeating shapes, the nested curves, the diagonal lines, and the way the pattern of light and dark enhance this.

The children are seeing patterns as a reduction, a simplification of the picture. There are also patterns in the content—fingers and their positioning represent a pattern, for example. The mimicking of the shapes of the fingers and those of the orchid and the stone in the ring (all things the children notice) are also elements of the pattern. Seeing this and communicating it comprise a reductive process. They require defining and categorizing the things seen. The purpose of this conversation, though, is to think about the meaning of the word *pattern*, and so I push the children to speculate on this. I want them to think about the nature of the patterns they have just described.

The children go on to discuss more patterns they can see in other things around the room, and finally they begin to debate the qualities of pattern (that they repeat, have regularities and a periodicity) and then begin to discuss if a pattern is a design.

FIGURE 35.2

Thomas: It's a design in the picture, well that wasn't just theirs, somebody designed it, they couldn't have just said, its light...dark, they had to make the picture, they had to design the picture, to make it, they had to make up the pattern.

Teacher: So a design is something somebody made?

Emily: They had to decide what they were going to do to make this and how they were going to make this, what they were going to do.

The children's descriptions of the patterns are derived from a reduction, a simplification of something seen, and they begin to recognize their own role in this. In thinking about science and about doing scientific research, we realize how important "seeing" patterns becomes. Patterns are in essence constructs that arise from descriptions. They are characterized by variables, constants, and operations. They make connections and enable us to see similarities, describe relationships, create regularities. They are made: created, imposed, and manipulated by people through selective vision. Seeing patterns is a simplification; it dichotomizes reality into the regular and the irregular.

MANIPULATING THE IMAGE ITSELF AND DISCOVERING THINGS

The third example concerns students in a graduate class we teach in which the focus is a comparison of art and science. The conversations described here concern the central project of the course, in which students are asked to develop over a fifteen-week period a portfolio in some artistic medium new to them but that they've "always wanted to learn." The purpose of this project is to reflect on the process of learning the "craft" or science behind an artistic medium and how this science, which is embedded in the art, enables self-expression and artistry. The class also visits a number of scientific laboratories in a parallel study of the art embedded in the work of scientists. A description of that facet of the class can be found in Osborne and Brady (2001).

Lora, a student in the class, is presenting her photography to the group. She has given us a short lecture on using a camera, making reference back to one of our readings from Ansel Adams's book *The Camera* (1980). Then she moves on to talking about darkroom techniques. She talks about the difficulties of obtaining an image like this one of a duck on water (see Figure 35.3).

The opacity of the backlit duck in contrast to the reflective and translucent water creates the pattern of lights and darks that would confuse a would-be predator. They also confuse the would-be photographer: how does one set the exposure for the brightness of the water and still image the details of the duck? Some of this can be dealt within the darkroom by selectively developing some parts of the exposure more than others. This however decreases the impact of the repeating patterns of light and dark—the confusion is *supposed* to be there, and increasing the ability for the viewer to "see" the information in the picture actually causes the story of the picture—a story contained in the confusion—to be lost.

In thinking about the technology behind photography, these technologies can be manipulated to enhance some things over others. The mindful use of these technologies enables the photographers to tell a story—their representation of what they saw is a story—relating much more to a viewer than

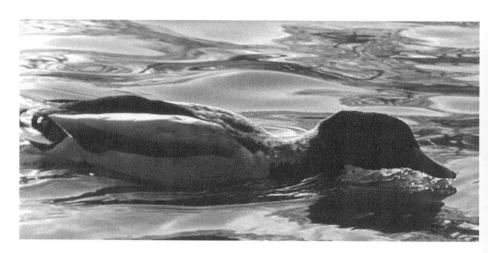

FIGURE 35.3

facts about the object imaged. Such a story unfolds to the viewer through questions: the viewer asks questions as he or she looks at the image, and the answers come from the viewer rather than the image stating them. Musing on the shape of the duck and the wavelets enriches our understandings of the morphology of ducks.

SEEING LEADS TO "THINKING ABOUT"

The final example describes how a visualization tool can enable a different perspective on something not previously thought about: through manipulating the tool, and the image, new perspectives occur. The act of manipulating the image (by manipulating the tool) distances one from the object being imaged and constructs a perspective enabling these ways of looking.

An activity we often do, in our classes with teachers and also with children when we go out into the community, is to let them play with miniature digital video cameras. Everyone, old and young, does two things with a camera within minutes of learning how to focus and move it about. They point it in their mouths and then in their ears, and they turn it back upon itself to record its own image on the monitors. What do they see when they do these things, and why is it so universal an activity? To look in their mouths and their ears, we can only think they are drawn to look at a part of themselves that others can view but they have never been able to see, repossessing something that is theirs but that has remained a secret to them *because they could not see it*. It is impossible to deny that we possess something in our culture by "seeing" it. Such an activity is reminiscent of the early days of the second wave of the woman's movement, when consciousness raising involved exploring our own bodies in ways that we had only allowed others to do before. In this instance, seeing is knowing.

The second thing everyone who handles these little video cameras does is hold the video camera up to its own image, resulting in a mysterious pattern of pictures of the camera spiraling smaller and smaller to a central point that the eye (or monitor) can't resolve. People peer at this central point, unblinking, looking for some mysterious secret. Now what are people searching for when they do this? We think that they seem to be searching for an essence, something inside the camera that goes beyond logic. Suddenly they are "thinking about" the camera as seeing through some sort of door into something they don't understand at all.

We return to two of our original questions "What is art?" and "What is science?" and pose a dichotomy: art is based on individual understanding but leads to universal truth. Science is based on universal truth but leads to individual understanding. In all the activities we have described, the art and the science are intertwined with each other, enable each other in both the doing and the production of understandings. That is clear, but what remain unaddressed are questions of what defines something as "art" rather than "science." What in essence is science or art? Our distinction that art leads to universal truths while science starts there seems thin when processes rather than products are examined.

Simplistic statements, such as "Science is objective reality, whereas art is subjective reality," imply observers that stand outside of an activity passing judgment rather than experiencing it or its meaning. It makes the designation *science* or *art* dependent on objectivity and commodification (Berger 1972). Such statements suggest not only that human activities create reality but also that reality somehow stands apart, as objective truth to which experience is anchored. In our classes, we use art (images)—doing/making art, looking at/ critiquing art—as a vehicle to see new things, enlarge the creative possibilities of doing basic science, and rethink the nature of doing science for our students. We use the making of art around scientific topics to enlarge observation of all domains—content, process, and subject matter—and to enable conversations about the nature of "seeing": what things are "seen" by science? What are "seen" by art?

REFERENCES

Adams, Ansel. 1980. *The Camera*. Boston: New York Graphic Society.

Bateson, Gregory. 1979. *Mind and Nature: A Necessary Unity (Advances in Systems Theory, Complexity, and the Human Sciences)*. New York: Hampton Press.

Berger, John. 1972. *Ways of Seeing*. New York: Penguin.

Eisner, Elliot W. 1998. Does Experience in the Arts Boost Academic Achievement? *Journal of Art & Design Education* 17:51–60.

Osborne, Margery D., and David J. Brady. 2000. A Comparative Exploration of Art and Science. *Arts & Learning* 16:15–31.

———. 2001. Imagining the New: Constructing a Space for Creativity in Science. In *Passion and Pedagogy: Relation, Creation, and Transformation in Teaching*, edited by Elijah Mirochnik and Debora C. Sherman, 317–32. New York: Peter Lang.

Weisskopf, Victor F. 1979. Art and Science. *American Scholar* 48:473–85.

Resources for Connecting Science and Literature

Barbara Tobin

Despite the challenging economic times, we are seeing an unprecedented burgeoning of well-written trade books that support an integrated study of science. The growing emphasis on expository writing in schools and the addition of science to the national assessments are helping fuel an increase in the quantity and quality of nonfiction science trade books being published.

THE ROLE OF TRADE BOOKS IN SCIENCE LEARNING

In a climate where the relentless march toward high-stakes testing of science achievement can narrow the curriculum to testable skills and stifle a teacher's flexibility and creativity, it is not surprising that many teachers have turned to trade books to enrich and extend the set curriculum and to get students excited about learning science. Trade books can play an important motivating role in supplementing science textbooks since they tend to be more colorful, engaging, and up-to-date.

Trade books are frequently used to introduce or reinforce a science concept, or to provide resources for student research across reading levels. With the increased availability of age-appropriate nonfiction texts, young readers now have the opportunity to read these books on their own instructional and independent reading levels.

Yet we must not fall into the trap of believing that science-based literature should be "used" only in the service of formal learning, as textbooks are. There are many enticing nonfiction books that can be enjoyed as pleasure reading for their own sakes as works of art and wonder. Such books can help break down the barrier between "school reading" and "real reading," and encourage students to independently develop and pursue their own interests in science. Surrounding children with a wealth of irresistible science trade books from an early age will help lay the foundation for a lifelong interest, even a career, in science. That students may do better on achievement tests because of the trade books is a short-term bonus.

There are cautions, however, in using trade books uncritically, especially in the younger grades where an integrated, literature-based approach is common. Studies about the use of trade books in science classrooms have shown that factual errors in these books can influence children's understanding of science concepts and lead to misconceptions, which can persist even in the face of contrary evidence.

It is therefore important to use high-quality trade books, ones that not only have been evaluated for their literary and artistic qualities, but also have been checked for accuracy of scientific content.

USING REVIEWS AND AWARDS TO IDENTIFY GOOD TRADE BOOKS

Perusing award-winning and notable book lists and reading book reviews from professional journals will uncover a wealth of well-written, accurate, and appealing books to nurture curiosity and support science learning. Much of this material is readily accessible online.

A quick way to access information about books mentioned in this chapter is to use online sites like amazon.com or bookfinder.us, where you can search by title, author, or keywords and be linked to selected reviews from major journals like *School Library Journal* and *Kirkus Reviews*. Reading multiple reviews provides varied perspectives to help you make choices.

For a far more powerful reference tool, try the Children's Literature Comprehensive Database, to which many library systems subscribe. This information retrieval service offers over 200,000 reviews of children's books from thirty-three review sources, including relevant science journals and organizations. It also includes book jacket images, book awards and best book lists, links to author websites, information about reading measurement programs, and curriculum tools that link to relevant lesson plans and teacher guides.

RAISING THE STATUS OF NONFICTION TRADE BOOKS

Science trade books have benefited greatly from the growing recognition and critical attention given to nonfiction, which has become an exciting field, as evidenced by the increasing number of nonfiction titles winning general literary awards, of new nonfiction awards, and of awards given specifically for science trade books.

In 1990 the National Council of Teachers of English set out to raise the balance of nonfiction titles in the major children's book awards by establishing the Orbis Pictus Award for Outstanding Nonfiction for Children. Many science titles have been honored, including Barbara Kerley's *The Dinosaurs of Waterhouse Hawkins* (2001) and Debbie Miller's *The Great Serum Race: Blazing the Iditarod Trail* (2003).

A decade later, the Association of Library Services to Children followed suit with the Robert Sibert Informational Book Award. Honored science titles include Sophie Webb's *My Season with Penguins: An Antarctic Journal* (2000) and Joan Dash's *The Longitude Prize* (2000).

Deserving science books are often spotlighted in the nonfiction category of such esteemed notable book lists as the *Bulletin of the Center for Children's Books'* Blue Ribbon lists.

SCIENCE BOOK AWARDS AND RECOMMENDED LISTS

Various professional science associations have developed their own awards and notable lists to recognize excellence in science trade books. Whilst retaining strict literary standards, the selection committees are more stringently focused on excellence in science writing and include more scientific expertise among the panelists.

Since 1973, the National Science Teachers Association (NSTA), in cooperation with the Children's Book Council, has produced an annual list of Outstanding Science Trade Books for Students K–12. The judges include science educators, coordinators, consultants, and teachers. The selected titles are published in an annotated bibliography, noting relevant *National Science Education Standards* (National Research Council 1996), in the March issue of the journal *Science and Children* and on NSTA's website.

Another excellent source is *SB&F* (*Science Books & Films*), a bimonthly critical review journal from the American Association for the Advancement of Science (AAAS). Since 1965, *SB&F* has exhaustively evaluated the merits (and demerits) of print and nonprint materials in all of the sciences and for all age groups. An extensive Best Books list appears in the January–February issue and on SB&F Online.

In 2005, *SB&F* launched its Key Awards for Excellence in Science Books to celebrate outstanding science writing and illustration. These awards are given to recently published books that promote scientific literacy and adhere to high standards of scientific accuracy; in addition, the inaugural awards conferred Lifetime Achievement Awards for significant and lasting contributions to children's and young adult science literature. The children's author winners were Patricia Lauber, Laurence Pringle, and Seymour Simon. Bernie Zubrowski won the hands-on science author award, and Jim Arnosky won the illustrator award. Nominees included Jean Craighead George, Kathryn Lasky, Vicki Cobb, and Aliki.

This lineup of notable science writers and artists is similar to the much longer and older list of the *Washington Post*–Children's Book Guild Award for Nonfiction, which also honors distinguished lifetime contributions. Other honorees here include Milton Meltzer, Gail Gibbons, and Steve Jenkins, who is noted for his visually stunning textured collages in books like *Actual Size* (2004).

There is a growing number of children's book awards in specialist science fields. The American Institute of Physics' Science Writing Awards recognize distinguished writing and illustration that contribute to the understanding and appreciation of physics and astronomy. The Giverny Award was established in 1998 by the 15° Lab for children's science picture books that teach young readers important scientific principles or encourage them toward science-related pursuits. The judges include experts in plant biology, biology education, geology education, and ecology.

EVALUATING SCIENCE TRADE BOOKS

The above resources offer guidance in selecting books deemed outstanding by specialists in both science and literature. Criteria for evaluating informational books can be adapted to suit your own purposes to sift through the hundreds of titles that never get an official seal of approval, yet may be equally worthy.

Specific criteria for the various science book awards can be found at the websites of the sponsoring bodies. The rigorous selection guidelines of the NSTA judges are a particularly useful model and relate to both content and presentation. Of critical importance is that information in the books be clear and accurate. Generalizations should be supported by facts, which should not be oversimplified to the point of being misleading. Significant personification and anthropomorphism should be avoided where it might create misconceptions.

Since the diverse panels of critics share a common core of evaluation criteria, it is not unusual for titles to appear on multiple lists when they meet the very highest standards in both literature and science. Examples are John Fleischman's *Phineas Gage: A Gruesome but True Story about Brain Science* (2002), which received numerous nominations for honors from the Orbis Pictus Award for outstanding nonfiction for children, the Bulletin Blue Ribbon, Dorothy Canfield Fisher, and James Madison Honor Books. Also Philip Hoose's *The Race to Save the Lord God Bird* (2004) was the recipient of the Boston-Horn Book Award and listed on the *Washington Post*'s Best Books for Younger Readers.

One of the most widely honored titles is Jim Murphy's *An American Plague: The True and Terrifying Story of the Yellow Fever Epidemic of 1793* (2003). This meticulously researched and documented account vividly traces the course of this medical and social calamity, whilst showing the struggles that medical scientists endure to overturn accepted yet insufficient beliefs of the time. Murphy leaves readers pondering the challenges of the evolution of drug-resistant viruses and also provides a wealth of sources for additional background reading, including Laurie Halse Anderson's novel that covers the same era, *Fever 1793* (2000).

There are many trade books that, although flawed from a strict informational perspective, may have other attributes that make them attractive to share. Some flaws can be compensated for by using them to teach students to become critical consumers. For example, in *My Light*, Molly Bang (2004) shares her passion for electricity with a fairly young audience. Her artistic talents have produced a picture book of great beauty likely to engage these readers. However, Bang's ambitious attempt to translate her copious research into a simple explanation of complex concepts about energy transfer is somewhat confounded by her poetic text, told from the perspective of the Sun itself: "I am your sun, a golden star. You see my radiance as light" (Bang 2004).

After children have enjoyed a first reading of this narrative at the aesthetic level, they can then discuss the Sun's personification and share some nonfiction books that present similar concepts in a different way, for example, Melvin Berger's *Switch On, Switch Off* (1990). Reading multiple sources helps them to verify evidence and become critical thinkers. Reading multiple genres helps them to master different types of writing. Their natural curiosity and love of playful language can be nurtured side by side. The rules against anthropomorphism are relaxed in poetry, for example the "concrete poems" of Joan Bransfield Graham's *Flicker Flash* (1999), which celebrates the energy and emotion of light in its many familiar forms.

LOOKING BEYOND NONFICTION TO SUPPORT SCIENCE LEARNING

Pairing nonfiction books with fiction and poetry can extend learning and pique interest by bringing new perspectives and showing the pervasiveness of science in many kinds of writing.

Poetry

As Mr. Newton, the caricatured science teacher in *Science Verse*, tells his class, "if you listen closely enough, you can hear the poetry of science in everything" (Scieszka and Smith 2004). Anything by the zany author-illustrator team of Jon Scieszka and Lane Smith is a tremendous hook for young learners, and this collection of parodies of traditional poems will have fans giggling before the book is even open. They are unlikely to learn much science here, though the periodic table on the endpages could invite discussion. Still, the book gives science a sense of humor. It is even hard to take offense at the ridiculous cartoon stereotype of Mr. Newton.

Distinguished poet and anthologist Lee Bennett Hopkins has put together a more thoughtful collection of poetry for young people in *Spectacular Science: A Book of Poems* (1999). New voices and old favorites like Carl Sandburg, Aileen Fisher, and David McCord celebrate the mysteries and glory of science. A poet's careful observations and search for analogies and patterns are at the heart of some important science explorations.

Valerie Worth's deceptively simple poems in *All the Small Poems and Fourteen More* encourage readers to look closely at common objects around them through fresh eyes and from unusual perspectives. A drifting soap bubble is compared to a hollow, revolving planet, "mapped with rainbows"; and a magnet trades "secrets with the North Pole" (Worth 1994).

In *Echoes for the Eye: Poems to Celebrate Patterns in Nature*, Barbara Esbensen and Helen Davie (1996) reveal a natural geometry in the repetition of shapes. Jagged lightning looks like the veins in our hands; and spirals echo from a nautilus shell, the Milky Way, and the Fibonacci patterns of a sunflower.

Paul Fleischman's *Joyful Noise: Poems for Two Voices* (1988) formats poems about insects for choral reading with voices that mesh like a musical duet. "Honeybees" contrasts the voices of a joyful queen and a disgruntled worker bee. After reading Sandra Markle's outstanding photo essay *Outside and Inside Killer Bees* (2004), students might be inspired to write and perform their own poems to show the differences between the two types of bees. An unusual perspective of the honeybee life cycle, as viewed through the lens of a worker bee, is provided by entomologist and cartoonist Jay Hosler's entertaining yet thought-provoking graphic novel, *Clan Apis* (2000).

There is a wealth of poetry available for spring-boarding into science themes. Christina Rossetti's short poem "Who Has Seen the Wind?" makes a perfect introduction to Vicki Cobb's nonfiction picture book *I Face the Wind* (2003), part of her excellent Science Play series for preschoolers. Cobb's exuberant, hands-on approach to learning science is reflected in a contemporary book design where playfully manipulated text helps create a synergy of words and pictures. A former science teacher, Cobb incorporates simple experiments in her books, which can be supplemented by the weekly experiments she posts on her website, Show-Biz Science (n.d.).

Graphic Novels

There is a growing recognition of the educational value of graphic novels to cater for multiple learning styles and to engage students who have grown up in a visually saturated culture. The Maryland Comic Book Initiative is the first statewide program for public schools to use a comics-based curriculum across the grades to supplement traditional reading materials. Such legitimization of a formerly spurned format will undoubtedly spur growth in this area. For younger readers, Robert Burleigh's short fictionalized biographies, *Amelia Earhart Free in the Skies* (2003) and *Into the Air: The Story of the Wright Brothers' First Flight* (2002), are part of the graphic novel series American Heroes.

A more complex interplay of text and visuals is evident in graphic novels for older readers, for example Jim Ottaviani's books about influential physicists and women scientists. His penetrating glimpses into the life of Niels Bohr, in *Suspended in Language* (2004) and *Two Fisted Science* (2001), could be compared with Naomi Pasachoff's biography from the Great Minds of Science series, *Niels Bohr: Physicist and Humanitarian* (2003), and Michael

Frayn's play *Copenhagen* (2000), which has been adapted into a television drama.

Memoir

Memoir is a developing genre for adolescent readers, told in a variety of narrative forms that bring a human face to science learning. Judd Winick uses the graphic novel format powerfully in *Pedro and Me: Friendship, Loss, and What I Learned* (2000) to tell the story of his roommate on MTV's *Real World San Francisco*, HIV-positive AIDS activist Pedro Zamora.

In another unforgettable memoir, *The Burn Journals*, Brent Runyon (2004) recounts with unflinching detail how, as an eighth grader on the emotional edge, he doused himself with gasoline and set fire to himself, sustaining burns to 85 percent of his body. In journal style, he takes us through the excruciatingly painful physical and psychological aftermath in a hospital burns unit and rehabilitation center.

Science Fiction

Good science fiction extrapolates imaginatively yet believably from our existing technological knowledge and scientific theories. The gap between the real and imagined uses of cloning is narrowing, and science fiction novels that creatively explore cloning and its moral and ethical issues from a variety of perspectives give young people the opportunity to grapple with many of the same issues that are part of our society's current discourse on human cloning (Crew 2004); they are, after all, already living in the imagined time of Kathryn Lasky's *Star Split*, the era of "Reproductive Reformation" (2001, 218). Lasky provides specific, accurate scientific details to help readers bridge that gap.

Part of a new trend for nonfiction books that examine the science in popular science fiction books is John and Mary Gribbin's *The Science of Philip Pullman's "His Dark Materials"* (2005), which explains in lay terms the various quantum physics theories used by Pullman in his controversial award-winning fantasy series (Pullman 2003). Roger Highfield has done something similar in *The Science of Harry Potter* (2002).

Weaving *Charlotte's Web*

A science inquiry can drive a thematic unit, which might begin with a nonfiction book like Darlyne Murawsi's *Spiders and Their Webs* (2004). A science inquiry might also arise spontaneously from a stimulus like the incredible photograph of dewdrops clinging to a spiderweb in Walter Wick's *A Drop of Water* (1997). A science inquiry can also be part of a unit of study built around a fiction book like E. B. White's *Charlotte's Web* (1952).

First graders love White's classic fantasy about a pig whose life is saved by the extraordinary web weavings of a spider called Charlotte. Fourth graders enjoy revisiting this book at a deeper level, using Peter Neumeyer's *The Annotated Charlotte's Web* (1994), which includes supplementary information such as White's early drafts, articles, and letters to his editor.

Before White started writing, he spent hours observing spiders in his own barn and read books about spiders to learn about their habits, capabilities, and temperament. He guarded against excessive personification of Charlotte, who teaches Wilbur the pig much about the true *Aranea cavatica*. Readers can examine reproductions of White's notes about spider anatomy, his sketch of the vectors of the web-spinning process, and notes he made about his spider protagonist to keep her true to the nature of her species.

White came to realize that "reality and fantasy make good bedfellows.... [T]here was no need to tamper in any way with the habits and characteristics of spiders" (Neumeyer 1994, 223). He said,

> [Y]ou should never lose sight of the fact that it [Charlotte's web] was a web spun by a true arachnid, not by a de facto person.... [S]piders do not talk to pigs, except in the world of fable. But when the conversation does finally take place, in that fabulous and pure world, it is indeed a spider who talks.... It is not a woman in spider's clothing. (224)

What a great way for young students to learn about the differences between and intersection of fiction and nonfiction writing. They could compare what they learned about spiders from *Charlotte's Web* with information gleaned from Melvin Berger's *Spinning Spiders* (2003), part of the Let's-Read-and-Find-Out science nonfiction series.

Older students may enjoy Ralph Fletcher's *Spider Boy* (1998), a realistic fiction story about a seventh-grade arachnophile and his pet tarantula. Despite the school bully's merciless taunting of "Spider boy from Illinois," Bobby continues to write in his spider journal, excerpts of which open each chapter with a wealth of scientific details about arachnids.

Intrigued readers could extend their learning in *The Tarantula Scientist* (Montgomery and Bishop 2004) from the excellent Scientists in the Field series. Naturalist Sy Montgomery joins photographer and biologist Nic Bishop as they follow spider scientist Sam Marshall through the dense rainforests of Guiana, then back to his extensive tarantula laboratory in Ohio. This is nonfiction science writing at its very best, and even the most reluctant of readers could not help but be compelled by Bishop's remarkable, sometimes rare photos.

Perhaps Sam Marshall was inspired to become an arachnologist by his childhood reading of *Charlotte's Web*. After all, science and literature are intricately interwoven.

REFERENCES

Anderson, Laurie Halse. 2000. *Fever 1793*. New York: Simon & Schuster Children's
    Publishing.

Bang, Molly. 2004. *My Light*. New York: Blue Sky Press.

Berger, Melvin. 1990. *Switch On, Switch Off*. New York: HarperTrophy.

———. 2003. *Spinning Spiders*. New York: HarperTrophy.

Burleigh, Robert. 2002. *Into the Air: The Story of the Wright Brothers' First Flight*. New York: Silver Whistle Paperbacks.

———. 2003. *Amelia Earhart Free in the Skies*. New York: Silver Whistle Paperbacks.

Cobb, Vicki. 2003. *I Face the Wind*. New York: HarperCollins.

Crew, Hilary. 2004. Not So Brave a World: The Representation of Human Cloning in Science Fiction for Young Adults. *The Lion and the Unicorn* 28 (2): 203–21.

Dash, Joan. 2000. *The Longitude Prize*. New York: Farrar, Straus & Giroux.

Esbensen, Barbara, and Helen Davie. 1996. *Echoes for the Eye: Poems to Celebrate Patterns in Nature*. New York: HarperCollins.

Fleischman, John. 2002. *Phineas Gage: A Gruesome but True Story about Brain Science*. Boston: Houghton Mifflin.

Fleischman, Paul. 1988. *Joyful Noise: Poems for Two Voices*. New York: Laura Geringer.

Fletcher, Ralph. 1998. *Spider Boy*. New York: Yearling.

Frayn, Michael. 2000. *Copenhagen*. New York: Anchor.

Graham, Joan Bransfield. 1999. *Flicker Flash*. Boston: Houghton Mifflin.

Gribbin, John, and Mary Gribbin. 2005. *The Science of Philip Pullman's "His Dark Materials."* New York: Knopf Books for Young Readers.

Highfield, Roger. 2002. *The Science of Harry Potter*. New York: Viking Adult.

Hoose, Philip. 2004. *The Race to Save the Lord God Bird*. New York: Farrar, Straus & Giroux.

Hopkins, Lee Bennett. 1999. *Spectacular Science: A Book of Poems*. New York: Simon & Schuster Children's Publishing.

Hosler, Jay. 2000. *Clan Apis*. Columbus, OH: Active Synapse.

Jenkins, Steve. 2004. *Actual Size*. Boston: Houghton Mifflin.

Kerley, Barbara. 2001. *The Dinosaurs of Waterhouse Hawkins*. New York: Scholastic Press.

Lasky, Kathryn. 2001. *Star Split*. New York: Hyperion.

Markle, Sandra. 2004. *Outside and Inside Killer Bees*. New York: Walker Books for Young Readers.

Miller, Debbie. 2003. *The Great Serum Race: Blazing the Iditarod Trail*. New York: Walker Books for Young Readers.

Montgomery, Sy, and Nic Bishop. 2004. *The Tarantula Scientist*. Boston: Houghton Mifflin.

Murawsi, Darlyne. 2004. *Spiders and Their Webs*. Washington, DC: National Geographic Children's Books.

Murphy, Jim. 2003. *An American Plague: The True and Terrifying Story of the Yellow Fever Epidemic of 1793*. New York: Clarion.

National Research Council. 1996. *National Science Education Standards*. Washington, DC: National Academy Press.

Neumeyer, Peter. 1994. *The Annotated Charlotte's Web*. New York: HarperCollins.

Ottaviani, Jim. 2001. *Two Fisted Science*. Ann Arbor, MI: G.T. Labs.

———. 2004. *Suspended in Language*. Ann Arbor, MI: G.T. Labs.

Pasachoff, Naomi. 2003. *Niels Bohr: Physicist and Humanitarian*. Berkeley Heights, NJ: Enslow Publishers.

Pullman, Philip. 2003. *His Dark Materials Trilogy* (The Golden Compass, The Subtle Knife, *and* The Amber Spyglass). New York: Laurel Leaf.

Runyon, Brent. 2004. *The Burn Journals*. New York: Knopf Books for Young Readers.

Scieszka, Jon, and Lane Smith. 2004. *Science Verse*. New York: Viking.

Show-Biz Science. N.d. Links to Vicki Cobb's Kids' Science Page. www.education world.com/a_lesson/archives/showbiz_science.shtml.

Webb, Sophie. 2000. *My Season with Penguins: An Antarctic Journal*. Boston: Houghton Mifflin.

White, E. B. 1952. *Charlotte's Web*. New York: HarperCollins Children's Books.

Wick, Walter. 1997. *A Drop of Water*. New York: Scholastic.

Winick, Judd. 2000. *Pedro and Me: Friendship, Loss, and What I Learned*. New York: Henry Holt.

Worth, Valerie. 1994. *All the Small Poems and Fourteen More*. New York: Farrar, Strauss & Giroux.

ADDITIONAL RESOURCES

Suggested Readings

Ansberry, Karen, and Emily Morgan. 2004. *Picture Perfect Science Lessons: Using Children's Books to Guide Inquiry, Grades 3–6*. Arlington, VA: NSTA Press. (Connects with *National Science Education Standards* [see National Research Council 1996, above].)

Butzow, Carol, and John Butzow. 2000. *Science through Children's Literature: An Integrated Approach*. Portsmouth, NH: Teacher Ideas Press.

Flagg, Ann, Mary Ory, and Teri Ory. 2002. *Teaching Science with Favorite Picture Books: Grades 1–3*. New York: Scholastic. (Connects with *National Science Education Standards*.)

Rice, Diana, Ann Dudley, and Christy Williams. 2001. How Do You Choose Science Trade Books? *Science and Children* (March): 18–22.

Websites

Book Links. Includes thematic bibliographies in curricular areas, discussion questions, and activities. Some science-related articles are archived: www.ala.org/ala/ProductsandPublications/periodicals/booklinks/booklinks.htm

Children's Literature. Links to major book awards (including science), themed reviews, teaching materials, and Children's Literature Comprehensive Database: www.childrenslit.com/award_link.html

Cooperative Children's Book Center. Links to major book awards and best-of-year lists, including "40 Books about Science and Scientists": www.education.wisc.edu/ccbc/

Lisa Bartle's Database of Award-Winning Children's Literature. Links to major book awards (including science): www.dawcl.com

37

The Place for Comics in Science Education

Shannon Casey

WHY COMICS FOR SCIENCE EDUCATION?

Leaf through any scientific periodical, and what do you find? Whether you've chosen the popular science newsmagazine *Discover* or an esoteric professional journal such as *Microbiological Reviews,* you will notice that most of the articles contain graphs, charts, diagrams, and illustrations. Scientists depend on graphics to augment, communicate, and organize their very complex ideas. Through graphic conventions, scientists communicate what they consider to be the most important data to support their ideas. And so do cartoonists. You may be surprised to know that comics and science share many key qualities that serve to make comics an effective, if unlikely, tool for the teaching and learning of science.

Behind any scientific discovery lies a dramatic tale of human enterprise and passion pitted against indiscriminate nature. While the everyday bench work done by cadres of lab scientists may not seem so glamorous, science does have all of the elements of a good story, or what Joseph Campbell calls "the hero's saga." Essentially, the protagonists set out to achieve a goal, then run into a mess of trouble and have to struggle through adverse conditions to finally achieve their goal. So too it is in science, where the drama of discovery unfolds for those in the laboratory or out in the field—far from the public's eye—until the cure is found or the existential mystery is solved.

In science fiction comics, lab scenes are mirrored by exciting midnight clashes between the lone scientist-as-superhero and his or her evil arch-nemesis. Think Peter Parker versus the Green Goblin, or Superman versus Lex Luthor. Can you recall what element renders Superman powerless? Or how Wonder Woman traveled around the globe? Odds are you can, because comics are memorable. Humans identify with comic book heroes. We get caught up in their iconic stories, and these myths become part of our shared culture. If such elements serve to make science fiction memorable, why not use them to serve science *nonfiction*, too?

Science fiction comics are, admittedly, more glamorous and dramatic than real-life science. But the exciting comic format can inspire young students to become scientists, or at least to be excited about communicating science textually and visually. Many adults who "do" science loved fantastical science fiction comics as adolescents. Adolescence is a time of identity development and increasing consciousness about the world at large. As teens forge themselves and grapple with the realities of life, they often seek role models who are powerful yet alternative to their parents or teachers. Similar to adolescence, comics are about genesis and transformation: from man to Superman, from powerless everyman Bruce Banner to the powerful Hulk. By personalizing and humanizing science through narrative and biography, students may identify with the protagonists and feel inspired by the content to become scientists.

Educators struggle to present science in a way that engages and challenges students without losing them. The sciences are rich with information and difficult to master, and so curricula and texts can seem inaccessible and static. Inventive teachers seek to enhance their curriculum with alternative sources such as recent journal articles, the science section of the newspaper, or popular science books. Science-based comics are another good choice for alternative content because, in comics, new vocabulary can be introduced within a well-developed context, and with supporting illustrations. Comics can simultaneously present a range of competing perspectives, voices, and ideas and prompt the processing and critical inquiry of these differing viewpoints. Also, like real-life science, the stories in comics are constantly evolving. Each installment in a comic series ends with a dramatic cliffhanger: will the hero overcome? Will the experiment succeed? Infusing science with drama can keep students connected.

WHAT IS A COMIC?

Simply put, a comic is a story told through a sequence of illustrated frames. Some comics are short and limited to a few pages or even just a few frames. Graphic novels, however, are longer, bound collections of chapters or previously released serial issues. In storytelling, the combination of words and graphics packs a significantly more powerful cognitive and emotive punch than text alone. Hackneyed as it sounds, a picture really is worth a thousand words! But please don't think of reading comics as facile. Comics use their own shorthand and require an esoteric literacy, just as mathematics and science do. Each illustrated frame is designed to affect an understanding by the reader, yet a lot of the narrative takes place *between* the frames and installments. The reader must use her or his imagination to connect what is happening both in and between the frames in order to render a unified whole in her or his mind's eye. To this end, the reader becomes a "silent accomplice" in constructing the story. Such participation is agency—an active, necessary, and powerful force in cognition. This specific skill set needed to read comics must be learned through exposure and used in conjunction with the text to construct the narrative. The reader must know that her or his eyes might travel in any direction on the page, depending on how the page layout was created to support the storyline. For teaching science, this ability to show movement can be critical to cognition. Think of describing Brownian movement or acceleration in text and equations alone versus *showing* them through a series of illustrated frames. Comics use sophisticated chronological cues, too, meaning they can effectively present phenomena in two, three, and four dimensions. Try doing that with simple text!

Comics also use emotional cues. Facial expressions indicating shock, happiness, or distress engage readers immediately (imagine eyebrows raised to show surprise). Another classic example of a comic convention is the "thought bubble," the most straightforward way to let the reader in on a character's inner dialogue and something cumbersome to convey in standard text. I mean, what is more interesting? Seeing the scientist scratching her chin and mulling competing hypotheses over in her mind or reading a static lab report?

An additional possible pedagogical benefit of comics is that pictorial narrative informs with less language, presenting the possibility of making scientific content more accessible to more people such as English language learners and learning disabled students. Students who love comics may get into the habit of picking up a book to engage themselves—and end up learning science in the process!

WHAT KINDS OF SCIENCE-BASED COMICS ARE BEING WRITTEN?

Comics traditionally and overwhelmingly deal with science fiction, but a handful of science-based comics exist and the corpus of work is growing. The three most notable authors currently creating science comics are Jim Ottaviani, Jay Hosler, and Larry Gonick. Ottaviani's graphic novels, *Two Fisted Science* (2001), *Dignifying Science: Stories about Women Scientists* (2003), and *Fallout* (2001), are biographies of important scientists and their discoveries, in which the main events and most of the dialogue are based on available evidence. In *Dignifying Science*, a different woman illustrator drew the biography of each woman scientist from Hedy Lamarr (did you know that?) to Barbara McClintock, emphasizing that women and men make important contributions to the writing and drawing of comics as well as to science. Dr. Jay Hosler, who is a Ph.D. biologist and teaches at Juniata College, is the creator of *Clan Apis* (Hosler 2000), which tells the life story of Nyuki, the honeybee, from a bee's perspective. Hosler has also written the *Sandwalk Adventure* series (Hosler 2003), wherein Darwin explains evolution to a mite living in his eyebrow. This fact-based and theoretically accurate graphic novel was, in part, a reaction to some educational comics out there that are decidedly *not* science, as they are about creationism. Hosler has a rich website in which he outlines the science associated with each story in *Clan Apis*. Recently, Ottaviani worked with Hosler and other artists to create *Suspended in Language*, a medley of comic styles narrating the life and discoveries of Niels Bohr. Larry Gonick's productions are less narrative and more content driven. He uses graphic conventions to illustrate scientific concepts and pair them with their proper equations. His *A Cartoon History of the Universe* is a classic that starts with the Big Bang and continues to tell the entire story of evolution. The prolific Gonick in collaboration with others has created more than fifteen "Cartoon History" graphic novels to explain such concepts as sex, chemistry, physics, and computers.

USING COMICS AS A METHOD OF ALTERNATIVE ASSESSMENT

If students enjoy and learn from the comic medium, why not ask them to communicate their acquired understandings by producing their own comics? Comics lend themselves very nicely to alternative assessment on several counts. First of all, comics are traditionally, and continue to be, the domain of the adolescent. Students are more likely to be agents of their learning if they feel they own the medium. Moreover, students can tap their own unlimited creativity to bring their learning to life, which is much more exciting than simply proving they have memorized facts by regurgitating them on a traditional test. Being able to write her own comic about a scientific idea requires a student to have a comprehensive understanding of the material in order to interpret and synthesize the knowledge, apply what she knows by constructing a narrative sequence, and explain what she has learned by illustrating the story elements and writing the text. The struggle of creating a good, narrative comic results in a level of understanding, ownership, and agency that rote memorization and repetition simply do not. Another benefit to having students create comics is that the artifact of a comic absolutely conquers the problem of plagiarism. Also, because comics accommodate explanatory drawings, curiosities, graphs, tables, and historical data, students can communicate multiple facets of their scientific knowledge and demonstrate their literacy through various devices.

Naturally, not every student is excited about the idea of creating his or her own comics; many students are insecure about their artistic ability. Try assigning the project to groups, where one or two artistic students may emerge. Also, not every unit in every discipline is well suited to being assessed with comics. Consider asking students to create the biography of a scientist, to dramatize a problem relevant to their lives, or to teach a scientific concept to the younger grades through a science-based comic.

As with any rigorous assessment, a comic takes a lot of thought and may need to go through a number of iterations to produce a final, grade-worthy artifact. The teacher should explain clearly to students what she expects, and scaffold the phases of development with discrete objectives at each checkpoint. Students can assess their peers using a teacher-made rubric or a rubric the class members develop for themselves. Each comic should include new vocabulary, accommodate any competing theories or viewpoints, and, ultimately, clearly, comprehensively, and creatively explain the scientific knowledge they have learned. Students often amaze us with their creativity when given half a chance, in which case the teacher can have their work bound and distributed to the next lower grade, to parents, and throughout the school community.

SOME ISSUES FOR USING COMICS IN THE SCIENCE CLASSROOM: A CONVERSATION WITH CATHERINE MILNE (NEW YORK UNIVERSITY) AND BARBARA TOBIN (CHILDREN'S LITERATURE CONSULTANT)

Shannon Casey laid out a thoughtful rationale for using comics in the science classroom as a learning tool and as a form of alternative assessment. Are there some implementation issues that teachers and parents need to be aware of?

Catherine: As I discussed the use of comics with you, Barbara, you made me aware of some issues I had not considered when thinking about using comic writing in a science classroom. My interest was in the possibilities comics present for learners to demonstrate that they really understand an area of science. Comics can involve learners synthesizing information from research that they have done on a science topic and recreating it in comic format. This involves learners in complex thinking skills as they incorporate scientific facts into a dramatic story. The comic becomes a true test of a learner's knowledge of an area of science.

Barbara: I like the approach you and Shannon have developed to use student-constructed comics to encourage higher order thinking skills, while providing a refreshingly creative, alternative assessment. To date, comics have mainly been used in classrooms to develop reading skills, to motivate reluctant readers, and, occasionally, to examine them as an alternative literary or artistic form. You open up new possibilities for extending a study of comics across the curriculum to content areas like science and social studies, thus facilitating more integrated learning and assessment. Creating science-based comics that meet both scientific criteria and those of a successful comic is a very labor-intensive physical, intellectual, and aesthetic task that will benefit from a multidisciplinary approach by teachers. Apart from clearly addressing scientific principles, these comics should also engage the intended reading audience in a motivating, interactive way that slabs of expository text cannot.

Catherine: Shannon encountered that challenge when she developed her own comic to explain how catalytic converters worked. She created an interesting storyline about a young woman who takes her car for emissions testing and is persuaded by her mechanic to have a new catalytic converter fitted. Whereas Shannon fulfilled the scientific criteria well, and her comic was for the most part very engaging, there were several large blocks of dense text in the middle, where the mechanic explained the intricacies of catalysis. That dampened your initial enthusiasm for Shannon's comic because it hindered your reading process. There must be a temptation for beginning comic writers to think that all information is important, but a reader can only understand so much at one time.

Barbara: Yes, Shannon's elaborate scientific explanation presented in static text, with no action and little graphic support, made me want to skip over that part. The narrative action slows right down, and we begin to lose the power of this medium to present science content in a clear, fresh way. If students are to be asked to create their own comics for credit, it is essential that they understand how this unique format works. Language arts teachers could help students learn the narrative conventions of comics, and art teachers the design principles. Analyzing some of the comics Shannon recommends above would lead students to determine the attributes of a successful comic book and how comics differ from more traditional ways of presenting information.

The challenge for students is to develop interesting characters to drive their story, to create a problem that builds tension, and to maintain momentum by not cramming too much information into one panel. Shannon needed to keep up the lively interaction between her two characters, and to continue drawing on her commendable toolkit of comic conventions, like thought bubbles and iconic images. For example, her playful use of a large, cloud-like font for the word *smog*, with ominous wavy lines emanating from it, brings an emotional depth to the narrator's explanation. Cartoonists create a synergy of text-illustration, where words become pictures, and pictures become words.

Catherine: I am reminded of Jay Hosler's *The Sandwalk Adventures*, in which he introduces us to the follicle mites, Mara and Willy. Hosler uses the interactions between Mara and Charles Darwin to present the theory of evolution to readers. Mara, Willy, and Darwin are the major characters that drive the story. There is a lot of science in the story, but character development is important because it leads us to care about what the main characters are thinking and saying. So the science message emerges from the interactions between characters we care about, even follicle mites!

Barbara: To help this happen, we should encourage learners to plan or storyboard their comics, beginning with their characters and setting, then developing the major plot points and interactions, which will lead into the specific dialogues that will propel their story from its beginning through the middle and on to a successful conclusion.

Students mustn't lose sight of the unique nature of the comic format by letting the burden of the scientific explanations be placed on textual delivery. We must remind them that much of the story can be carried pictorially, and yet we don't need to be great artists to represent important objects and concepts. As Scott McCloud points out when discussing iconic abstraction in comics, "[S]imple elements can combine in complex ways, as atoms become molecules and molecules become life" (McCloud 1993, 45). Like the atom, he says, great power can be locked in a few simple lines, and released by the reader's mind; a simple style doesn't necessarily mean a simple story.

Catherine: Your suggestions also remind me of another issue that learners often find difficult about science. Science is experienced through the use of the senses. Yet the explanations often require the use of models and theories that are not directly observable. For example, you can experience directly chemical reactions such as teeth whitening using something like Crest White Strips. All you need to do is look in the mirror. However, explaining how White Strips cause that chemical reaction requires the use of atoms and molecules, which is not a sensory experience. Some of your suggested strategies provide a means for presenting explanations without slowing down the narrative.

Barbara: Perhaps the use of fantasy fiction, as opposed to realistic fiction, lends itself better to explaining more abstract phenomena. Joanna Cole and Bruce Degen use a hybrid genre blend to help younger children understand less tangible science concepts in their Magic School Bus series (e.g., *The Magic School Bus Inside a Hurricane* 1995). Although cartoon format is only one part of their nonlinear, multiple-perspectives approach to telling a science story, prospective comic writers might benefit from examining the ways these authors break science content up into accessible parts using speech balloons, labels, wall charts, and "factoids" listed on simulated student notebook pages. Students who have grown up in this digital age are well used to processing text in this hypertext-like format, where they control the flow and direction of the reading.

I am optimistic that given the exposure and opportunities in science classrooms such as we have discussed here, students will not only feel more excited about learning science and gain deeper understandings, but some students may be inspired to build a lasting interest, even a career, in contributing their own hand-crafted comics to the small but growing field of science-based comics. Jay Hosler, the very talented entomologist-cartoonist, is an incredible role model!

REFERENCES

Cole, Joanna, and Bruce Degen. 1995. *The Magic School Bus Inside a Hurricane*. New York: Scholastic Inc.

Gonick, Larry, and Craig Criddle. 2005. *The Cartoon Guide to Chemistry*. New York: HarperCollins.

Hosler, Jay. 2000. *Clan Apis*. Columbus, OH: Active Synapse.

———. 2003. *The Sandwalk Adventures*. Columbus, OH: Active Synapse.

McCloud, Scott. 1993. *Understanding Comics: The Invisible Art*. New York: Harper-Collins.

Ottaviani, Jim. 2001. *Two Fisted Science*. 2nd ed. Ann Arbor, MI: General Textronics Labs.
————. 2001. *Fallout*. Ann Arbor, MI: General Textronics Labs.
————. 2003. *Dignifying Science: Stories about Women Scientists*. Ann Arbor, MI: General Textronics Labs.

WEBSITE

In his website, Jay Hosler outlines some of the stories and the supporting science, especially *Clan Apis*: www.jayhosler.com

38

Using Driving Questions to Motivate and Sustain Student Interest in Learning Science

Joseph Krajcik and Rachel Mamlok-Naaman

INTRODUCTION

How can we capture and then sustain student interest in science? What can we do to help ensure students will find what they learn in science useful? Advances in understanding how students learn science have led to new approaches to science instruction. The Highly Interactive Computers for Education (hi-c^{3}e) group at the University of Michigan has expanded and articulated our understanding of project-based learning (PBL). Phyllis Blumenfeld and colleagues (1991) describe PBL as an approach to teaching and learning rooted in inquiry pedagogy that is consistent with social constructivist ideas. Five prominent learning features characterize this model of instruction:

a. *active construction*, which states that students must be actively engaged in explaining, generalizing, hypothesizing, representing, and so on in order for deep understanding to develop;

b. *situated cognition*, which states that learning occurs best where contextualized in situations meaningful to the students;

c. learning occurs through repeated exposure to the practices of a *community* of practitioners;

d. the communities of practitioners constitute communities of a disciplinary *discourse*; and

e. *cognitive tools* can expand what students can learn.

Hice has elaborated two related versions of PBL, project-based science (PBS) and design-based science (DBS). Joseph Krajcik, Charlene Czerniak, and Carl Berger (2003) have extensively described PBS, and David Fortus and colleagues (2004) have worked out the details of DBS. The philosophy of PBS is that students need opportunities to construct knowledge by solving problems through asking and refining questions; designing and conducting investigations; gathering, analyzing, and interpreting information and data; and developing explanations, making conclusions, and reporting findings. PBS takes an investigation approach to the teaching and learning of science. In PBS, students find solutions to meaningful questions through investigations, collaboration, and the use of learning technologies. The core principle underlying PBS is to create meaningful context in which students can find value for engaging in the learning of science through an investigative process.

DBS focuses on the need to design artifacts that are related to real-world problems and are intended to be interesting and challenging to the students. The goal of DBS is not to instruct the students about design, but rather to engage in design in order to learn science. Therefore, to enable students to engage in design without having to teach them about it, Rachel Mamlok and colleagues (2001) suggest organizing each unit around a learning cycle in which students actively engage in *constructing* knowledge by asking and refining questions; developing personal ideas; designing and building models; conducting investigations; gathering, analyzing, and interpreting information and data; drawing conclusions; and reporting on their findings.

Several common characteristics between PBS and DBS are that each (1) centers around authentic tasks for sustained periods of time, (2) leads to the creation of artifacts by the students, (3) encourages the use of alternative assessment, (4) utilizes learning technologies, (5) builds upon student collaboration, and (6) views the teacher as a facilitator and a learner along with the students rather than as a source of knowledge.

Phyllis Blumenfeld and colleagues (2000) made this comment about student motivation in project-based learning environments:

> It is insufficient to provide opportunities designed to promote knowledge that is integrative, dynamic, and generative if students will not invest the effort necessary to acquire information. (375)

The novel and varied activities found in the PBS and DBS instructional materials may also provide opportunities to motivate student learning. However, exposure to interesting content, phenomena, and laboratory activities alone does not determine whether measured learning will take place. The students need to find the tasks to be worthwhile, and they need to make the effort to learn the science content to see any changes between pre- and posttest scores.

Joseph Krajcik, Charlene Czerniak, and Carl Berger (2003) describe the driving question as the central organizing feature found in PBL. A driving question is a well-designed question that students and teachers elaborate, explore, and answer throughout a project. The driving question is the first step in meeting all of the other key features because the question sets the stage for all activities and investigations. Both DBS and PBS require a question that is meaningful and important to learners and that serves to organize and drive activities. As students collaboratively pursue answers to the driving question, they develop understandings of key scientific concepts associated with the project.

In this chapter, we use several examples of materials that have either a PBS or a DBS framework, that students found meaningful, and that have been used by numerous teachers.[1] We structured the units around driving questions chosen to be interesting and challenging to students.

WHY USE A DRIVING QUESTION?

Traditionally, curriculum developers organized science textbooks around topics, such as planetary motion, ecology, chemical reaction, force and motion, the human body, and so on. Often, materials will mention reasons for learning the topic, but these ideas are only briefly mentioned and do not provide compelling reasons for studying the topic. Although knowing such reasons might initially increase interest, the reasons are not the driving forces behind learning, and teachers frequently find it difficult to hold students' attention when focusing on topics that a textbook suggests as important. In reviewing middle school curriculum materials, Sofia Kesidou and Jo Ellen Roseman (2002) found that unit openers that included questions framed only a small part of the unit, making it unlikely that students will stay engaged throughout the unit. Most materials do not have a driving question or other strategies that provide and maintain a sense of purpose throughout the unit.

Often in schools, students see what they are learning as separate from their own interests and lives, and ideas are taught in isolation from each other without an organizer that can pull together the overall ideas. By using driving questions, students can see the purpose for learning, and when they learn a new skill or concept, they can immediately apply it to help answer the driving question and subquestions.

A driving question provides a need to know and leads to students doing inquiry to find solutions to the question. A driving question has value if teachers can link together various components of a project, if it links content and process together, and if it connects various subject areas so that students (1) develop understandings about the connections, (2) find it meaningful, (3) engage in an intellectual problem over time, and (4) see how the science they are learning applies to their lives.

Driving questions help teachers link the various concepts taught, which helps students develop integrated understandings that are linked together rather than isolated, making the ideas students learn usable. Students find the questions and, thus, learning relevant to their lives and their world or culture, and they find the projects exciting and interesting. The questions become anchors for tying together all the new information that students are learning, which results in the information making more sense. For example, in the unit "What Is the Quality of Air in My Community?" designed for seventh graders, the learners develop an understanding of the factors that affect air quality with a focus on the particulate nature of matter and chemical and physics properties. Learners examine different sources of pollution in their neighborhood and use archived data to compare air quality in Detroit with that of other cities. Jackson, Krajcik, and Soloway (1999) developed a learner-centered modeling tool that students use to build, test, and evaluate qualitative models of air quality. Students also use a learner center 3-D molecular visualization and assembly tool, to build representations of complex molecules.

FEATURES OF DRIVING QUESTIONS

Krajcik, Czerniak, and Berger (2003) describe several key features to a good driving question:

- Feasibility: students should be able to design and perform investigations to answer the question.
- Worth: questions should deal with rich science content and process that match district curriculum standards.
- Contextualization: questions should be anchored in the lives of learners and deal with important, real-world questions.
- Meaning: questions should be interesting and exciting to learners.
- Ethical: questions should not harm living organisms or the environment.
- Sustainability: questions should sustain students' interest for weeks.

A key feature of a good driving question is feasibility. A driving question is feasible if (1) students can design and perform investigations to answer it, (2) resources and materials are available for teacher and students to perform the investigations necessary to answer it, (3) it is developmentally appropriate for students, and (4) it can be broken down into smaller questions that students can ask and answer. Rivet and Krajcik (2004) designed the middle school PBS unit "How Do Machines Help Me Build Big Things?" to help learners meet several district, state, and national standards including balanced and unbalanced forces and their effect on motion, simple machines and how they work together in complex machines, and the concept of mechanical advantage. The driving question is broken down into four subquestions that help to structure and organize an eight-week project. Each question gives students an opportunity to engage in various investigations. For example, in order to answer the subquestion "How do I move things?" students need to learn the concepts of force and motion and to use process skills such as predicting, observing, explaining, and sharing definitions.

Students work in collaborative groups to design and conduct an investigation to explore how they might change a simple machine to increase its mechanical advantage. Then they apply the knowledge gained in their investigation to designing their own machines. A ninth-grade DBS unit developed by Fortus and colleagues (2004), "How Do I Design a Structure for Extreme Environments?" sets the goal of designing and building a model house that can withstand extreme environmental conditions. This unit is composed of five learning cycles, dealing with weather conditions, technical drawings, different sources of loads, shape and structural integrity, and thermal insulation. In each cycle, students construct and modify model houses based on the knowledge constructed in that and the previous cycles.

Another key feature of the driving question is its worth. A worthwhile question contains rich science content that students can explore and that meet district, state, or national standards. Perhaps the most important feature of a worthwhile driving question is the quality of science content and process that it can encompass. Worthwhile driving questions are directly linked to learning goals in such a way that they cannot be answered without students gaining an understanding of the intended science content. "What Is the Water Like in My River?" a middle school PBS unit, helps students meet learning goals about water ecology and basic chemical principles. Seventh-grade learners construct an integrated understanding of science concepts such as watersheds, erosion, and deposition, and chemistry concepts such as pH and dissolved oxygen. Students use electronic sensors and graphing software (i.e., pH, dissolved oxygen, and temperature probes) to collect and visualize real-time data as they conduct water quality testing. Students use a learner-centered modeling tool, developed by Jackson and colleagues at the University of Michigan (1999) to represent their understanding of the watershed, erosion and deposition, runoff, and the impact of these factors on water quality.

Contextualization is a key feature of good driving questions. A contextualized question anchors the project in an important real-world situation that has important consequences. Students may not see immediately how a question relates to the real world or perceive its consequence. However, a good driving question presents the opportunity to draw students in and teaches them to see how it is related to their lives. In the PBS learning unit "Why Do I Need to Wear a Helmet When I Ride My Bike?" (aimed at eighth graders), students focus on the investigation of the physics of collisions. Learners develop an integrated understanding of force, velocity, acceleration, and Newton's first law in the context of being pitched off their bike, getting injured, and learning how helmets work. Learners also develop strategies for interpreting and visualizing physical phenomena graphically. The project integrates the use of computer-based labs such as the use of motion probes to explore the relationship in distance and time graphs. Students develop an appreciation of the driving question by watching a video of a teenage boy who talks about head injuries he sustained in a bicycle accident because he did not wear a bicycle helmet. They also observe a demonstration of an unprotected egg rolling down a ramp in a cart, representing a student riding a bicycle without a helmet. When the cart reached the end of the ramp, it hit a barrier that caused the egg to fly out of the cart and break when it landed on the table. Students revisited the egg and cart demo several times throughout the project. Each time, students responses to what occurred became more sophisticated. The video and the egg and cart demonstration were common experiences shared by all students that helped them see how the driving question was related to their lives.

Students should find driving questions meaningful to their lives. Usually, meaningful questions are those that students see as important and as interesting to them. Meaningful questions intersect with their lives, reality, and culture. The DBS unit developed by David Fortus and colleagues (2004), "How Do I Design a Battery That Is Safer for the Environment?" has a goal of designing and building a wet cell that makes use of nontoxic materials. This unit has four cycles addressing toxic materials and their disposal, different types of batteries, materials they are made from and the health hazards related to these materials, how batteries decay and how decay is measured, and electric circuits and electrochemistry. Since the students had little understanding at the start of the unit of how an electrochemical cell is constructed, they designed, constructed, and tested liquid batteries and created posters of their batteries at the end of the unit. The posters and their justifications show that the students had constructed an understanding of the unit's concepts, and how common issues of science and technology apply in a real-world context. In this unit, students on one hand get acquainted with the utilization of the batteries in different domains, including the human body, and on the other hand, they learn about the problems of the disposal of batteries, and they better understand that materials from human societies affect the physical and chemical environment.

An important feature of a good driving question is that it is ethical. As students investigate their questions, they should not harm living organisms or the environment. Further, as students explore and find solutions to questions, the procedures they use should be ethical. There should be no leeway in this position. A driving question is ethical if it holds paramount the safety, health, and welfare of living things and the environment. Students sometimes have questions that are of interest to them but, if investigated, would harm living things or the environment.

A final feature of a good driving question is that it sustains student engagement over time. Teachers and students can work on a good question for weeks or even months. In addition to holding student engagement over time, a sustainable driving question encourages students to study information in great detail. In the DBS unit developed by David Fortus and colleagues (2004), "How Do I Design a Cellular Phone That Is Safer to Use?" the goal is to design a cellular phone that minimizes potential radiation and sound hazards without compromising customer appeal. This unit contains five cycles dealing with the potential hazards of electromagnetic (EM) radiation, the historical development of form and function in telephone technology, general wave characteristics, sound waves, and EM waves. The cell phone curriculum begins by giving the students design specifications that indicate the requirements that their cell phones will have to meet. Having the students watch a video of an ABC 20/20 story on cell phone safety sets the context of cell phones as potential sources of ill health and environmental hazards. The students then read a newspaper article on the same subject, and a classroom discussion is held about the facts and opinions expressed in the video and in the newspaper article. Next, the students engaged in a series of units that are organized around the different stages of the design process. They completed web-based research on the different components of a safer cell phone (environmentally friendly batteries, safe sound levels, and EM radiation levels). To summarize the changes in the form and function that telephones have undergone since their invention, they examined a museum collection. The students collected data from teenagers about their daily use of phones and summarized this in a report. Laboratories were available where students could build a battery and study battery chemistry, build a speaker and study sound waves, and learn about EM radiation from a series of demonstrations using TV antennas and a broadcast signal. Throughout these activities, the students participated in the design studio process by utilizing cycles of production and reflection. Students express their ideas through drawings and written essays, and then get feedback about their ideas in pin-up and critique sessions. The variety and meaningfulness of the driving question and various tasks allowed students to stay engaged in this project for six weeks.

HOW DO WE COME UP WITH DRIVING QUESTIONS?

In identifying driving questions, a teacher must first clearly know what the learning goals of the unit are. Developing a driving question that students might be interested in but that does not align with learning goals will lead to teachers abandoning the use of the driving question. Krajcik, Czernaik, and Berger (2003) identity several ways in which a teacher can develop driving questions:

- Personal experience and interests. Reading, method classes, hobbies, teacher workshops, and interaction with students during teaching may serve as good sources for driving questions. For instance, recycling may be a personal interest to some, and it could also serve as the basis for a driving question. A driving question such as "Where does all our garbage go?" should help students meet learning goals related to ecology and environmental science.

- School curriculum. Brainstorming driving questions from curriculum is critical, since the assessment of students is based on particular content, and the teacher can ensure that important outcomes are met. A driving question such as "Why do I need to wear a helmet when I ride my bike?" should fit into a curriculum dealing with Newton's first law of motion, velocity, acceleration, force, and the relationships between them.

- Listening to students. Driving questions developed from listening to students and using their experience are potentially more meaningful to students. Some teachers create "test banks" of students' ideas, in which they record their questions. For example, students may raise questions related to air quality because many students either suffer from or know others who have asthma. Other students are interested in how diseases spread from one individual to another.

- Media. Driving questions can also result from reading newspapers, listening to the news, watching documentaries and other shows on television, and surfing the web. For example, the driving question "How can I design a cellular phone that is safer to us?" came up after watching a documentary regarding the hazards of cellular phones and reading a similar article in the newspaper.

- Listening to other teachers. The personal experiences of colleagues are good sources of driving questions. Teachers who are interested in nutrition, for example, may inspire their colleagues with ideas related to this topic. Such projects should match learning goals related to biology.

- Published curriculum materials. Published curriculum includes textbook series, computer programs, and CD-ROM applications. These published curriculum materials often provide a structure for selecting driving questions that meet curriculum standards, and contain ideas for activities, resources, and background information about the science content. For instance, most high school chemistry books have a chapter on isomers, but most students don't often see why they should learn about isomers. However, if you show students two different substances that have very different properties but the same molecular weight and composition, you have a chance of hooking their interest in the importance of molecular structure.

However, it is important to emphasize that the driving questions need to meet important learning goals. Teachers always need to keep in mind their learning goals. They can help students develop driving questions by (1) creating rich classroom environments, (2) helping students generate subquestions to the main driving question, and (3) linking the driving questions to students' prior experiences.

HOW DOES A TEACHER USE A DRIVING QUESTION THROUGH A UNIT?

Once a teacher has chosen a driving question to guide instruction, he must keep referring to the driving question. It does little good to just initially mention the driving question and then not refer back to it. The power of the driving question lies in it being used as a linking device or connector throughout a project. The driving question gives a focus to the project's tasks and to what students are learning. By keeping the driving question in mind as teachers lead students through different activities, they will help the students contextualize what they are learning and help maintain continuity throughout the project. The driving question can be an excellent vehicle to maintain focus and support integration, but only if the teacher takes an active role in using the driving question throughout a project.

Making connections to the driving question are not always obvious and often pose challenges for teachers. Students seldom see how principles and tasks are related to the driving question until the teacher takes the time to help students see the links. There are several techniques for applying the driving question throughout a project.

One excellent and simple technique is to refer frequently to the driving question. At the end of every class and activity, before beginning each new activity, and periodically throughout the day, the teacher can ask, "Why are we doing all this?" She can ask questions like "We are learning about sound waves; how is that related to how a cell phone works?" A driving question only serves as a connector if the teacher helps students see the connections.

Another technique is to develop ways to link activities and investigations in class to the driving question. Teachers need to consider how to introduce a project's driving question so that the connections between the question and classroom activity start out firmly planted in students' minds. Throughout the project, teachers can reinforce this connection by asking periodically, "Why is this investigation important to our project?" and "How does it help us answer our driving question?" Students will come to expect teachers to ask these questions and will develop good responses.

Jonathan Singer and colleagues (2000) suggest that another way of linking the content and the class activities to the driving question is to have the driving question visible at all times through the use of a driving question board in the classroom. The teacher can point to the question every time she asks, "Why are we doing this?" Students will soon learn to look at the wall or bulletin board where the question is posted to help them stay on track.

Another technique is to have students periodically write in their journals about how what they are doing is related to the driving question. Some teachers use this technique in conjunction with class discussion that encourages students to tell how what they do in class on a given day is related to their driving question. Because these projects are long term, the potential for losing focus is high, so continual reminders of the driving question can prevent fragmented learning and keep students from getting lost.

CONCLUDING COMMENTS

Driving questions can help students see the value and meaning of what they learn. However, if such environments do not help students learn, then there is little need for the pain and challenge of changing our teaching. A variety of studies conducted by educational researchers—including Marx and colleagues (2004), Schneider and colleagues (2001), Tinker and Krajcik (2001), and Williams and Linn (2003)—have examined student learning in project-based environments. Their work illustrates that students can meet important learning goals in PBL environments. Taken as a whole, these findings demonstrate that carefully designed, developed, and enacted project-based learning environments that make use of the driving question can lead to students experiencing substantial learning gains.

However, as Blumenfeld and colleagues (2000) discuss, successful enactment of project-based learning brings many challenges for teachers. These challenges result because PBL depends on the content and pedagogical knowledge of teachers as well as their practical skills. To successfully enact projects, teachers need (1) new teaching strategies, (2) new classroom management skills, (3) deeper understanding of content, and (4) pedagogical content knowledge. Using projects and driving questions holds much promise for promoting engagement and student learning, but this will only occur with sustained professional development.

NOTES

The research reported here was supported in part by the Center for Curriculum Materials in Science (NSF-ESI-0227557). Any opinions expressed in this work are those of the authors and do not necessarily represent either those of the funding agency, Weizmann Institute of Science, or the University of Michigan. Joe completed this work at the Weizmann Institute of Science as the Weston Visiting Professor of Science Education. The authors contributed equally to this manuscript and are listed in alphabetical order.

1. Throughout this chapter, we make reference to several curriculum units. You can learn more about these units at www.hice.org/know.

REFERENCES

Blumenfeld, Phyllis C., Barry Fishman, Joseph S. Krajcik, Ronald W. Marx, and Elliot Soloway. 2000. Creating Usable Innovations in Systemic Reform: Scaling-UP Technology-Embedded Project Based Science in Urban Schools. *Educational Psychologist* 35:149–64.

Blumenfeld, Phyllis C., Elliot Soloway, Ronald W. Marx, Joseph S. Krajcik, Mark Guzdial, and Annemarie S Palincsar. 1991. Motivating Project-Based Learning. *Educational Psychologist* 26:369–98.

Fortus, David R., Charles Dershimer, Joseph S. Krajcik, Ronald W. Marx, and Rachel Mamlok-Naaman. 2004. Design-Based Science (DBS) and Student Learning. *Journal of Research in Science Teaching* 41:1081–110.

Jackson, Stephen, Joseph S. Krajcik, and Elliot Soloway. 1999. *Model-It: A Design Retrospective, Advanced Designs for the Technologies of Learning: Innovations in Science and Mathematics.* Hillsdale, NJ: Erlbaum.

Kesidou, Sofia, and Jo Ellen Roseman. 2002. How Well Do Middle School Science Programs Measure Up? Findings from Project 2061's Curriculum Review. *Journal of Research in Science Teaching* 39:522–49.

Krajcik, Joseph S., Charlene M. Czerniak, and Carl Berger. 2003. *Teaching Children Science in Elementary and Middle School Classrooms: A Project-Based Approach.* New York: McGraw-Hill.

Mamlok, Rachel R., Charles Dershimer, David Fortus, Joseph S. Krajcik, and Ronald W. Marx. 2001. A Case Study of the Development of a Design-Based Curriculum. Paper presented at 2001 NARST Annual Meeting, Liberating Minds through Disciplined Inquiry; Liberating Inquiry through Disciplined Minds, St. Louis, MO, April.

Marx, Ronald W., Phyllis C. Blumenfeld, Joseph S. Krajcik, Barry Fishman, Elliot Soloway, Robert Geier, and Revital T. Tal. 2004. Inquiry-Based Science in Middle Grades: Assessment of Learning in Urban Systemic Reform. *Journal of Research in Science Teaching* 41:1063–80.

Rivet, Ann E., and Joseph S. Krajcik. 2004. Project-Based Science Curricula: Achieving National Standards in Urban Systemic Reform. Paper presented at the Annual Meeting of the National Association for Research in Science Teaching, New Orleans, LA, April.

Schneider, Rebecca M., Joseph S. Krajcik, Ronald W. Marx, and Elliot Soloway. 2001. Performance of Students in Project-Based Science Classrooms on a National Measure of Science Achievement. *Journal of Research in Science Teaching* 38:821–42.

Singer, Jonathan, Ronald W. Marx, J. Krajcik, and Juanita Clay Chambers. 2000. Constructing Extended Inquiry Projects: Curriculum Materials for Science Education Reform. *Educational Psychologist* 35:165–78.

Tinker, Robert, and Joseph S. Krajcik, eds. 2001. *Portable Technologies: Science Learning in Context. Innovations in Science Education and Technology.* New York: Kluwer Academic/Plenum Publishers.

Williams, Michelle, and Marcia Linn. 2003. WISE Inquiry in Fifth Grade Biology. *Research in Science Education* 32:415–36.

39

Students as Curriculum Guides: Environmental Science

Cassondra Giombetti

Science is powerful as it inherently connects a wide range of disciplines, raises multitudes of questions, and inspires creative thinking toward real-world problems and issues. It is relevant to our lives, is interesting, and allows for creativity and for extensive arrays of interests. It is a field that science educators often express passion for, not just an obligation to teach. Yet each day, these same teachers are confronted with students who claim that science is *not* relevant and *not* interesting. How, then, are these passionate teachers to pass on, or at the very least share, this enthusiasm to generations of students?

How to best engage and interest students in science is a large and complicated question, one that can in no way be answered in one book chapter. However, it is one that can be discussed and considered within the field of environmental science by looking both to the subject matter and to the students. This chapter seeks to consider how student interest and student voice can be tapped as part of the curriculum development process specifically within environmental science classrooms. While many (if not most) of the ideas discussed here can be applied in different classrooms and amongst different science disciplines, this chapter will look to environmental science to introduce and ground these ideas here and now.

"SCIENCE IS NOT RELEVANT TO MY LIFE"

Science as a field is an ever-changing discipline. New advances and discoveries are being made every day. Some innovations are very small, such as minute video cameras used by doctors to look inside our bodies; other advances take us far from earth to the moons of planets thousands of light years away from our own. These advances, while amazing, often feel removed from our own lives; we are grateful that those small video cameras exist, but most of us will not likely need or use one in our lives. Likewise, stars and moons are beautiful aspects of the night sky, but few humans are likely to actually visit them. Then there are other advances that are physically located within our own personal experiences and yet remain removed and foreign as to *feel* as though they are light years away (e.g., agricultural science and meat processing). As a result, many students view the sciences incorrectly as something not relevant to their own lives. While each science discipline varies in ease of connecting science to student life, environmental science is one discipline that can be more easily made directly relevant to students' lives and experiences.

WHAT IS ENVIRONMENTAL SCIENCE?

Environmental science concerns relations between organisms and their environment, is interdisciplinary, and supports and is supported through nearly all other scientific disciplines, including chemistry, geology, and physics *as well as* social science disciplines such as environmental law and economics. Environmental science does not consider one specific aspect or component of the living world; rather, this approach considers the earth as a whole, one that is integrated and unavoidably interconnected. The relations and interrelatedness of the multitude of components of earth are as important as these components individually. In the classroom, the relationships between components, including humans, guide how and what is taught.

A Brief History

In the past, environmental science, and environmentalism in general, has taken various forms. Early environmentalists and nature writers such as Henry David Thoreau observed and wrote about the beauty and intricacies of nature (e.g., *Walden; or, Life in the Woods*; Thoreau 1995). Aldo Leopold, in his book *Sand County Almanac*, described the elaborate relationships he observed on his farm and muses that "there are two spiritual dangers in not owning a farm. One is the danger of supposing that breakfast comes from the grocery, and the other that heat comes from the furnace" (1966, 6).[1] John Muir hiked, camped in, and fought for the existence of the redwood forests in Northern California, forests that today are known as Muir Woods and remain protected as a national monument. The seminal work, though, is *Silent Spring*, written by Rachel Carson and published in 1962. Carson moves beyond description and advocacy to point fingers at the actions of human beings as a direct cause for loss of life and environmental degradation. Specifically, she describes the use of pesticides and the potential for environmental fallout that are still only beginning to be fully understood. *Silent Spring* marks the swing into awareness around human activity and excess that environmentalists are still fighting against today.

Today environmentalism is manifested in many ways. Universities offer undergraduate and graduate degrees in environmental science, where students studying this discipline can focus on topics ranging from the chemistry of soils to the politics of agriculture. Governmental agencies, existing on both the state and the national levels, work to acknowledge and support environmental issues (e.g., nationally, the U.S. Environmental Protection Agency [EPA]; or regionally, the Pennsylvania's Department of Environmental Protection [Pa DEP]) through physical resources, financial support, and education. Today's environmentalists include lawyers, scientists, and passionate laypeople alongside reflective nature writers and activists. All are interested in making a positive difference in the history and health of our planet.

Bringing Environmental Science into the Classroom

The wide variety of categories and roles described by terms like *environmentalism* can be daunting when these subjects are considered for inclusion in a classroom curriculum. The fact that environmental science is based in a foundation of knowledge that is grounded in the earth itself suggests that teaching environmental science requires engaging in a *very* large and complex foundation of knowledge. But return to the definition for a moment, "the branch of biology concerned with the relations between organisms and their environment," and consider the words *relations between*. These relationships are an integral aspect of environmental science. It is not simply the knowledge of *how* the different individual components of earth function, but also how these multitudes of components relate to each other and to their environment that becomes a key focus. Further, as humans are one of these components, how *we* relate to the other components of the earth is as important as how trees, other mammals, and insects relate to each other.

Thus, two main tenets are integral in the teaching and learning of environmental science. The first is the existence and importance of the interconnectedness of the earth, and the second is the place of humans in this complex system. Humans (our physiology, habits, and basic needs) are as much an integral part of this larger system as the plants and animals that are often the focus of science classes. The fundamental difference, though, is that humans maintain the ability to choose our actions as we are aware of what we do on a conscious level as compared to other organisms. Environmental science, then, looks to elevating students' awareness around their actions and encouraging a role as steward.

STEWARDSHIP BEGINS WITH AWARENESS

How are teachers today to impart the importance of stewardship upon generations of students who feel that science is not relevant to their lives? The first step toward this role of stewardship is through awareness bringing us full-circle back to the classroom and issues around student disinterest in science. Creating a space for student interest in the environmental science classroom allows for the inclusion of student-identified topics as a means both to provide relevant science curriculum *and* to support and encourage this role as steward of the earth. As students engage in material they have participated in identifying as interesting and relevant, opportunities for linking science to student lives are increased.

Obviously, there are many different approaches to identifying and creating an environmental science curriculum that is specific to a given classroom, school, and community. This chapter does not seek to define the ultimate curriculum; rather, it considers a potential approach toward this end. These ideas are grounded in the two assumptions introduced earlier and look heavily to the student as co-creator or guide during the curriculum development process. By infusing student-identified interests into the curriculum, the curriculum then becomes relevant to the student, thus smoothing the path for the teacher to impart the importance of interrelatedness and stewardship.

GUIDED FIRST BY THE STUDENTS

One option for approaching the development or identification of an environmental science curriculum for a specific classroom is to look to the students in that classroom as guides. Allowing the students to identify the focus of the curriculum creates a space for their interests and questions as a means to guide the direction of the class. This is not an easy and simple task, and it requires both the teacher's own dedication to the process and ample flexibility. What may look on day 1 to be a lesson about trees may morph into a lesson about deforestation by day 6 and further into a community education plan by day 15. What is important, here, is that the teacher is aware of and prepared for these twists and turns.

This does not require that the teacher "know everything" about any given topic, but rather asks that the teacher be devoted to supporting the direction of the class along the path of the students' choosing. Assigning Internet research may pique student interest at first, but without proper and sufficient scaffolding, students are likely to lapse into retrieving personal e-mails, reading sports scores. or playing games. This approach does not, by any stretch, imply that the student is allowed "free rein" of the classroom and subject; rather, the teacher taps into the extensive interests, energy, and creativity that students carry with them to school every day.

This approach to curriculum design is likely to develop around student-identified questions or concerns. What is relevant to students in their lives and in their classrooms is often centered around questions growing out of their own personal experiences. For instance, streams are a rich site for supporting student interest. Students of all ages enjoy playing in the water, plants, and animals living in and around streams are interesting and engaging, and most students have spent at least some time in contact with some aspect of flowing water (culverts, streams, rivers, etc.). Engaging students in dialogue around water allows the teacher to listen for any questions specific to water in general, but more specifically for any "wonderings" around streams in their community. If the students display interests specifically in the organisms living in the stream, the teacher may direct a unit on stream ecology. If the students express concern over the amount of pollution or lack of vegetation around a particular stream in the community, the teacher may suggest a unit around community education, including learning opportunities about both the stream ecosystem and stewardship. In both of these examples, students are exposed to the relationships between the stream and organisms living in and around the stream and how human actions impact this particular ecosystem. Both of these options also allow for various multidisciplinary activities, including social studies as students engage with their community, literacy skills as students read science texts and write science reports, and communication skills through engaging with the community around promotions of awareness and advocacy. Both examples also allow for changes in the direction of the course as new questions and topics are identified along the way.

LEAVING ROOM FOR STANDARDS

A classroom looking primarily to the student for guidance is likely to take different directions for each topic identified. While this is a very engaging way to experience science in the classroom, today's educational climate requires that educators must also contend with guidelines and expectations specific to the curriculum and subject material. In 2002, Pennsylvania passed standards for environment and ecology in public schools (Pennsylvania Department of Education 2002). Like so many of the standards in place today, this document categorizes specific content areas that are identified as key points around which student understanding should be developed. Each area is broken down into four or five main skills or concepts that describe, through three or four bulleted points, what students should know or be able to do by fourth, seventh, tenth, and twelfth grades. Teachers in Pennsylvania public schools *must* consider these standards as guides for environmental science and ecology classes, but the standards do not need to be viewed only as restrictive. Rather, there is room for variation within each guideline.

For instance, consider the standards for agriculture and society at the seventh-grade level. The first concept looks like this:

Pennsylvania's public schools shall teach, challenge, and support every student to realize his or her maximum potential and to acquire the knowledge and skills needed to

  a. Explain society's standard of living in relation to agriculture.
  1. Compare and contrast agricultural changes that have been made to meet society's needs.
  2. Compare and contrast how animals and plants affect agricultural systems.

3. Compare several technological advancements and their effect(s) on the historical growth of agriculture.

4. Compare different environmental conditions related to agricultural production, cost and quality of the product (Pennsylvania Department of Education 2002, 10).

This standard supports both the interrelatedness and stewardship foundations that are key to environmental science, and also allows for specific student interest to guide the direction of the curriculum. According to this standard, students must build basic biological and ecological understandings around agriculture (i.e., life cycles and basic needs of products such as wheat, soy, and fruit) *and* understand how humans are related to and impact these processes. But where the teacher chooses to place the lesson in relation to the students' own lives is flexible. If a particular classroom identifies questions around where the cafeteria food originates, the teacher may direct the unit toward a specific food group. If another classroom is located in a community where small family farms are being bought and developed into housing tracts, that teacher may guide the lessons toward a community and civic focus alongside the basic environmental and ecological understandings related to farming.

CONTEXTUALLY RELEVANT AND INTERDISCIPLINARY

The descriptions included here did little to define the context of the classroom. This was, on some level, intentional, as it is impossible to know who the readers of this chapter will be. But it is worth noting that what has been included here applies to any number of contexts. Where it is often "easier" to relate to the environmental context of rural or even suburban settings because trees, birds, grass, forests, and fields literally surround the students, urban contexts are also varied and rich in environmental resources. The interdisciplinary tendency of Environmental Science allows for the needs, interests, and details specific to each context to shine. Three examples are as follows:

- *Rural* students might identify the agriculture ecosystem as one worth knowing, and the curriculum may find it includes water quality issues designed around runoff polluted by fertilizers and animal waste, or land preservation issues as housing tracts are built upon previously existing small family farms.

- *Suburban* students might identify a poorly managed local park that once was known for fine fishing and design a curriculum that includes the biology of the stream and organisms living therein as well as community advocacy designed to educate the local community about the importance of such a park.

- *Urban* students may raise questions around vacant lots where the curriculum may include lessons around ecosystems and the varied life (plants, insects, and animals) that exists within a given lot as well as the opportunity to design and organize a new use for the vacant plot of land, such as an urban tree farm that can provide local residents with saplings to plant in their own yards or in other vacant lots.

In each setting, the students' interests and concerns can guide the direction of the class and a curriculum that focuses on concepts such as ecosystems, the water cycle, erosion and runoff, and the connections of the environment to humans and human actions. In each setting, the curriculum works to impart science concepts that are applicable to many other situations (e.g., chemistry, biology, policy, and social studies) and embraces both assumptions integral to environmental science: relationships and stewardship. The key in each case is the link back to the students' interests and questions. Attempting to relate the issues around agriculture to students whose life experiences are specific to the inner city will likely bring the teacher back to the challenge of "science is not relevant to my life." But looking to a vacant lot as a source of information to teach similar issues opens many doors.

There is no one best way to teach environmental science. Much of how one chooses to approach the curriculum depends on the personal strengths and comfort zone of the teacher as well as standards and other requirements placed on the curriculum by the school or district. But the vast foundation of knowledge within the discipline of environmental science becomes a strength by allowing for flexibility and opportunities to identify the most interesting, relevant, or important topics as guided by local circumstances and opportunities.

NOTE

1. Aldo Leopold, *Sand County Almanac* (New York: Ballantine Books, 1966).

REFERENCES

Carson, Rachel. 1962. *Silent Spring*. New York: Houghton Mifflin.
Leopold, Aldo. 1966. *Sand County Almanac*. New York: Ballantine Books.
Pennsylvania Department of Education. 2002. Academic Standards for Environment and Ecology. www.pde.state.pa.us/k12/lib/k12/envec.pdf.
Thoreau, Henry David. 1995. *Walden; or, Life in the Woods*. New York: Dover.

ADDITIONAL RESOURCES

Suggested Readings

Bouillion, Lisa M., and Louis M. Gomez. 2001. "Connecting School and Community with Science Learning: Real World Problems and School-Community Partnerships as Contextual Scaffolds." *Journal of Research in Science Teaching* 38 (8): 878–98.
Fusco, Dana. 2001. "Creating Relevant Science through Urban Planning and Gardening." *Journal of Research in Science Teaching* 38 (8): 860–77.

Russell, Helen R. 1998. *Ten Minute Field Trips*. Arlington, VA: National Science
    Teacher Association.

Websites

The North American Association for Environmental Education is a national organi-
    zation supporting environmental education through web resources including
    event calendars, grant and funding opportunities and job postings, an annual
    conference and a large publications database: http://naaee.org
Many national governmental agencies offer free online resources for educators, in-
    cluding the following:
U.S. Environmental Protection Agency (EPA), Office of Environmental Education:
    www.epa.gov/enviroed/
U.S. Department of Agriculture: www.usda.gov
U.S. Geological Survey (USGS): www.usgs.gov/education/
Individual state governmental agencies also offer resources for educators, including
    guest teachers, physical resources, and field trips. In Pennsylvania, some
    examples are the following:
PA Department of Conservation and Natural Resources (DCNR): www.dcnr.state.pa
    .us/education/
PA Fish and Boat Commission: www.fish.state.pa.us
PA Game Commission: www.pgc.state.pa.us/education/index.asp

40

Student Interest–Focused Curricula

Gale Seiler

Curriculum usually refers to the subject matter or content that a teacher addresses in a particular class. However, a curriculum also goes beyond content to include the entire plan of instruction that details what a teacher will teach and what students are to know, how they are to learn it, what the teacher's role is, and the context in which learning and teaching will take place. This holistic view is the starting point from which we will consider what a student interest–focused view of curriculum contributes to science education and how it can complement the current standards-based focus.

Most commonly, when educators talk about making the science curriculum "interesting" to students, they attempt to arouse the students' curiosity, hold their attention, or increase their involvement in the lesson. Efforts to interest students in these ways often include the use of hands-on activities and supplemental materials such as videos. These activities and materials are usually chosen by the teacher or suggested or mandated by the curriculum or textbook developers. Thus, although they may be aimed at developing or tapping into student interest, they are not responsive to or emergent from student interests. This is an important distinction that Dewey made in saying that education ought to take into account students' interests (1938).

Such efforts to arouse student interest assume that the students have few or weak interests of their own, or that interests that they do have are not appropriate for science class, and that it is the job of the curriculum and the teacher to create interest in science. However, as we go about our daily lives, we come to know our physical and natural world; in doing so, we develop interests in it, and these provide important links with science content. Interest can be thought of as a state of curiosity, concern, or attention to something. Thus students might have an interest in music, or in playing a particular sport, or in lizards or video games. Children have an array of interests from their lifeworlds, ranging from nascent to deep, and they bring these with them when they walk into the science classroom.

A student interest–focused curriculum does not mean simply a student-centered approach aimed at getting students more involved in science class. Rather, it recognizes that students already have interests in many science-related topics; thus, the focus shifts from designing a curriculum that creates student interest to one that recognizes ongoing interests and uses them to build understanding, motivation, and participation in science. It provides a place for students themselves to influence the curriculum, as the curriculum adapts or is adapted to their suggestions, questions, curiosities, pastimes, and passions. Approaches to this type of student responsiveness may include creating opportunities for students to suggest, request, or design their own science activities (or even their own science course) or the use (both sponta-neously and in planning) of questions and examples offered from the stu-dents' passions and experiences. We will refer to these as *direct* and *indirect input* into student interest–focused curricula.

Much of the research into this type of student-emergent curricula has been done in the context of low-achieving, African American, urban schools (Seiler 2001), since these are sites where student interest is often disregarded or considered nonexistent. The students in these schools are often envisioned as being disinterested in science. However, my belief in the power of student interest in science topics was strengthened in these schools as I listened to students who were considered apathetic to and disinterested in science as they struggled to connect textbook science with experiences and interests from their own lives. Audio and videotaping of student interactions in urban sci-ence classrooms have yielded a rich array of linkages that students make with science *on their own!* It was in inner-city high schools that the power of student emergent connections with the science curriculum was revealed.

OPTIONS FOR CURRICULUM DESIGN

Direct Student Input

Direct student input into the science curriculum occurs when students themselves make the "important decisions about what to study, how to study it, and how to demonstrate to others what they [have] learned" (Grace 1999, 50). When students function as curricular informants, they shape the cur-riculum and have the power to align it with their interests.

There are several approaches where science input can emerge directly from students. These include situations in which students are specifically asked what they want to learn, to design lessons for themselves or others, or to examine enacted lessons and generate suggestions for changes in subsequent lessons.

In an informal science lunch group, I asked a group of African American teens to talk about their interests outside of school and if there was any science in them (Seiler 2001). Week after week, they took on the challenge, finding science in their hobbies, passions, and curiosities such as drumming, video games, wrestling on television, and sports. Each week's activities were jointly planned based on what we talked about the previous weeks. Using balloons to explore Newton's laws led a student to ask, "Can you bring in some helium balloons?" Encounters with helium helped us understand characteristics of gases related to density and how these affect the vocal chords and hitting a baseball in Denver. Two of the young men were talented drummers, and under their guidance we investigated how attributes of a drum affect its sound, the physics of sound, and connected this to the vocal chords and the eardrum.

Thus, one way to generate a student interest–focused science curriculum is to ask the students to be involved in its development: "What are you interested in, and is there science in it?" or "What do you want to learn in science?" This approach was employed in a biology elective class at the same high school (Seiler 2002). The goal of this course was to elicit ideas from the students about topics they wanted to learn in human anatomy. While discussing body systems in the first three class periods, the students brought up at least thirty-five topics or questions. These represented an array of topics including menstrual cramps and PMS, diabetes, asthma, sports injuries such as those sustained by Allen Iverson (a basketball player for the Philadelphia 76ers), lead poisoning, and urine drug tests.

In cogenerative dialogue sessions that follow science lessons, a teacher and student representatives talk about the preceding lesson in order to design changes in the learning environment and curriculum that can be implemented to improve teaching and learning (LaVan and Beers 2005). These sessions provide a successful means for incorporating student voice in understanding what is happening in a classroom and for students together with teachers to generate plans for subsequent classroom activities and actions. These approaches, along with examples in the following section, correspond to Noddings's view of an interactive curriculum that grows from interactions between teachers and students (1995).

Indirect Student Input

It is also possible for teachers to use more indirect student input to develop student interest–focused curricula. Teachers who are attentive listeners, and are able to recognize and tease out student ideas, questions, and interests during discussions and within the doing of science, can adapt the curriculum to respond, either immediately in the same class period or very soon thereafter. This often necessitates retuning a teacher's ear, perhaps to the students' vernacular way of speaking, and retuning a teacher's toolkit not to be immediately dismissive of student interruptions and student-to-student cross talk. They can be rich sources of student interest.

Many student connections with science take the form of questions that a teacher might consider offhand or even off-task, but they represent significant intellectual efforts by students to connect science with their lives and their experiences. During a class discussion on the excretory system, a student posed the question of why urine appears to have an elevated temperature when a person is suffering from a head cold or infection. The class considered this question as a hypothesis, and with little prompting suggested a way to test its viability by urinating in a cup and measuring the temperature with a thermometer obtained from the school nurse. The eagerness with which the students engaged in this impromptu experimental design session is indicative of the power of switching from teacher input to student input in designing laboratory activities.

Analogies play an important role in science education, and many traditional analogies grace our science textbooks and classrooms. A cell is like a factory. DNA is like a twisted ladder. But what about student-generated analogies? An adrenaline rush is like Popeye on spinach. The autonomic nervous system is like a brain on layaway. Such student-generated analogies can be part of a curriculum open to student interests and voice. Similar to textbook and teacher-generated comparisons, these student analogies provide a bridge between current knowledge and the new knowledge or target concept. Offered by students, they can be taken up by teachers and students, and used to cogenerate understanding. Their power lies in the capacity of the teacher to create room for them in the curriculum, to use them, and to allow them to be used by the students, including the recognition of the limits of the usefulness of each analogy.

Relying on familiarity with students—their likes and dislikes; their favorite movies, sports teams, and players; and their families' and communities' strengths and concerns—teachers can anticipate student interests, connections, and questions. This is what Delpit (2002) describes in the use of hair products and hair chemistry to connect a middle school curriculum with the interests of African American students. We have used similar approaches for several years and have found that efforts to incorporate student interests into long-range planning and curriculum design become more successful when the teacher is familiar with the students' culture and lifeworlds. Teachers who continually learn from and about their students can use that information in focusing curricula in subsequent semesters.

A high school biology curriculum for an urban charter school represents an attempt to recognize student interests and provide opportunities for voice and choice while addressing state life science standards. Each unit includes a driving question (e.g., What's wrong with Alonzo Mourning, and why can't he play basketball?), and all content is approached through attempts to answer these questions, which are rooted in student interest in popular culture or community issues. From the driving questions, students generate subsidiary questions or topics: "I think we need to know about urine" or "What does the kidney do?" New questions can be added throughout the unit, as new awareness and connections arise. In understanding Alonzo Mourning's kidney condition, it was not initially apparent to the students that they would need to understand the roles of osmosis, diffusion, and active transport in the cells of excretory system. Student inquiry groups similar to those described by LaVan (LaVan and Beers 2005) are a key component in how the curriculum is enacted as students learn to formulate questions and use evidence in the practice of science. As the unit unfolds, the teacher selects from an array of activities, investigations, data explorations, readings, and assessments included in the curriculum, but the choice and sequence of these can be varied in response to the subsidiary questions identified by the students. The initial generation of the curriculum stemmed from several years of close study of student interests and work with students in curriculum development; the driving questions are intended to be reconsidered each year in relation to changing student interests.

EMOTIONAL ENERGY AND STUDENT INTERESTS

Traditional science curricula tend to lead to teacher-directed science instruction in which interactions follow an initiation-response-evaluation (IRE) pattern described by Lemke (1990) and there is little student to student crosstalk. However, discourse from classrooms with curricular attention to or space for student interests differs in this regard. Frequently, there are long sequences of cross-talk uninterrupted by teacher talk. Here students build on each other's ideas, often mirroring, repeating, or completing each other's statements, and are more willing to attempt the appropriation of scientific discourse (Seiler 2002).

In the biology elective class, the students repeatedly asked to do dissections and suggested organs and organisms to dissect. The practices of the school and classroom provided constraints that initially prevented dissections from being done. However, when specimens were eventually made available for dissection, videotapes showed that these class sessions were unique in several ways. When allowed to pursue their own interests, students participated more and stayed involved for lengthy periods of time as they exhibited creative practices in doing science. In addition, there was evidence of greater cooperation and solidarity among the students. Over and over, we have found that there are marked differences in engagement when student interests are addressed and/or students are given opportunities to express their choice regarding what or how to learn (Carambo 2005).

The examples provided here have alluded to the power of a student-emergent curricula in increasing student engagement and motivation in science class, but have not suggested why this might be so. The reasons are complex and related to the social, cultural, and historical contexts in which the teaching and learning are taking place. The concept of emotional energy (Collins 2004) provides an understanding of why student interest–focused curricula is often linked with different types of classroom cultures that more successfully engage students in learning and doing science. They provide opportunities for students to achieve synchrony, solidarity, and positive emotional energy that lead to agency and purpose.

IMPLICATIONS OF STUDENT INTEREST–FOCUSED CURRICULA

We have seen that a variety of approaches by teachers can be successful in allowing curricular interests to emerge from the students. Teachers, who are knowledgeable and experienced, with solid science content knowledge and a rich repertoire of teaching strategies, are often able to be responsive to students in the ways described here, and to work both within the moment as well as in the long term to identify student interests and adapt the curriculum to them. However, less science-oriented teachers are also able to be true facilitators and to provide wider opportunities for student ownership of topics aligned with their interests.

In recent years, an increasing emphasis on standardized science curricula has had the effect of diminishing opportunities for teachers to generate local curricula connected to the students and their communities and lives. Perhaps unnecessarily, school personnel often feel that they no longer have the discretion to mold the curriculum as they have in the past, to make it responsive to and appropriate for their student population. Instead, lessons are based on a standardized science curriculum written with much attention to what scientists and educators think is important, and much less to what students think. In the past, teachers were more connected with the communities in which they taught, and it was more likely that they would have been aware of student interests. Before the era of specified content and measurable learning outcomes, teachers would have felt that they had more latitude in making choices of content and approach. However, the previously described biology curriculum, designed around driving questions connected to the students' interest in popular culture and sports, and issues in their communities, shows that a focus on student interests and attention to standards are not mutually exclusive.

Contrary to the common lament, curricular responsiveness is even more important now in light of declining enrollment in upper-level science courses. The content of most science curricula does little to generate student enthusiasm and interest, and it is generally recognized that, particularly in science, student interest and curiosity that are strong in the lower grades decline during schooling. Attention to student voice and choice in science is particularly important for African American and Latino students. Their persistent achievement gap in science and lack of a presence in the professions of science are a civil rights issue (Moses 2001) that demands access to science that connects with their interests, combined with diligence not to confine students to their current and prior experiences and not to stereotype them in terms of their interests. Student relevance in the form of connections between science topics and student interests does not have to attend only to pre-existing interests of students. As positive emotional energy is created through interest-adapted curricula, further possibilities appear for the teacher and curriculum to mediate between the students and topics that are not of significant interest in their lives.

In a recursive fashion, curricula that are connected to student interests require and also generate alternative ways of thinking about and approaching science; this is an important strength of this approach. For example, when teachers create opportunities for the curriculum to respond to students' voice or choice, it fosters students who see science and themselves in new ways and participate differently in science class; and when they participate in new ways, they seize and create space for their interests to be voiced. It is neither possible nor necessary to say which comes first, since they fuel each other. As described by Barton (1999), reflexive science both affords and is afforded by responsive, adaptive teaching.

A final thought lingers on the need to be aware of the students' position in all of this. Students are accustomed to being neither asked what they want to learn in science nor expected to overtly connect science with their lives. I have found that it takes time for students, particularly students who have been racially or culturally marginalized by society and schooling, to accept the notion that topics of interest to them can even be considered science. With time and effort to enact a curriculum that sustains positive emotional energy, the recursive process described above can occur.

REFERENCES

Barton, Angela C. 1999. Learning to Teach a Multicultural Science Education Through Service Learning: A Case Study. *Journal of Teacher Education* 50: 303–12.

Carambo, Cristobal. 2005. Learning Science and the Centrality of Student Participation. In *Improving Urban Science Education: New Roles for Teachers, Students and Researchers*, edited by Kenneth Tobin, Rowhea Elmesky, and Gale Seiler, 167–84. Boulder, CO: Rowman & Littlefield.

Collins, Randall. 2004. *Interaction Ritual Chains*. Princeton, NJ: Princeton University Press.

Delpit, Lisa, and Joanne K. Dowdy, eds. 2002. *The Skin That We Speak: Thoughts on Language and Culture in the Classroom*. New York: New Press.

Dewey, John. 1938. *Experience and Education*. New York: Macmillan.

Grace, Marsha. 1999. When Students Create Curriculum. *Educational Leadership* (November): 49–52.

LaVan, Sarah-Kate, and Jennifer Beers. 2005. The Role of Cogenerative Dialogue in Learning to Teach and Transforming Learning Environments. In *Improving Urban Science Education: New Roles for Teachers, Students and Researchers*, edited by Kenneth Tobin, Rowhea Elmesky, and Gale Seiler, 149–65. Boulder, CO: Rowman & Littlefield.

Lemke, Jay. 1990. *Talking Science: Language, Learning, and Values*. Norwood, NJ: Ablex Publishing Corporation.

Moses, Robert P. 2001. *Radical Equations: Math Literacy and Civil Rights*. Boston: Beacon Press.

Noddings, Nel. 1995. *Philosophy of Education*. Boulder, CO: Westview Press.

Seiler, Gale. 2001. Reversing the "Standard" Direction: Science Emerging from the Lives of African American Students. *Journal of Research in Science Teaching* 38 (special issue on urban education): 1000–14.

———. 2002. Understanding Social Reproduction: The Recursive Nature of Structure and Agency within a Science Class. Ph.D. diss., University of Pennsylvania.

PART 7

RESOURCES FOR TEACHING AND LEARNING SCIENCE

41

The Internet and Science Education

Karen Elinich

Science learning takes place wherever someone stops to wonder. If a student brings imagination and curiosity to daily living, the moments and opportunities for wondering are boundless. Science educators hope that students can recognize the multitude of sites that become settings for science education—and that only a few of them are inside the walls of schools.

In recent years, the Internet has become one such setting. Learners of all ages are using the Internet—and the World Wide Web in particular—to develop their own understanding of science and technology. Both formal and informal science learning institutions now recognize this reality and are strategically using the Web as a venue for improving the quality of science learning.

THE NATURE OF SCIENCE

Fundamentally, the Web offers empowerment. Not so long ago, science happened entirely behind closed doors. K–8 students memorized some basic facts from the history of science. High school students recreated some well-known experiments—safe ones with certain outcomes. Undergraduates might be allowed to sneak a peek through a window in the door, and graduate-level students might be welcome to knock. Only after many years of study would a novice be welcome to cross the threshold and actually enter the world of science.

In many ways, the Web has unlocked and thrown open those doors so that students of all ages can encounter images of what it means to do science. As soon as the Web began to gain a foothold, collaborative science projects began to engage students with the real practice of science. The JASON Project, the GLOBE Project, and Journey North are just a few of the early efforts that used the Web to engage K–12 students in ongoing, real investigations that called for students to use the real skills of science—including observation, data analysis, and communication with colleagues. Suddenly, K–12 teachers had become empowered to create classroom learning environments where students were challenged to wonder.

The JASON Project predates the Web. Founded in 1989 by Dr. Robert Ballard, the well-known underwater explorer who located the wreck site of the *Titanic*, the JASON Project seeks to involve students in the real work of science research. When the Web became a presence in classrooms and homes, the JASON Project immediately established a Web presence for itself. The exact nature of the science work that students do varies from year to year, just as the real work of Dr. Ballard and his team varies. Where Dr. Ballard goes, the JASON Project and its "Argonauts" follow. Today, JASON is one of the best-established and longest-lived science learning adventures on the Web.

The Global Learning and Observations to Benefit the Environment (GLOBE) Program is an international environmental science research and education program, and is one of the Web's best examples of a collaborative science learning environment. GLOBE is a cooperative effort of schools, led in the United States by a federal interagency program supported by NASA, the National Science Foundation, and the U.S. State Department, in partnership with colleges and universities, state and local school systems, and nongovernment organizations. Internationally, GLOBE is a partnership between the United States and over 100 other countries. Students around the world take scientifically valid measurements related to the atmosphere, water, soil, and land cover. They use the Web to collect data, and then work with peer students and scientists to interpret the aggregated data from all over the world. Since the program began in 1995, K–12 students have collected, logged, and analyzed millions of measurements. The students who participate in the GLOBE program are developing their scientific thinking skills. They are also demonstrating that excellent science learning can happen beyond the classroom and through the Web.

Journey North is a global study of wildlife migration and seasonal change. Thousands of students across the country track animal migration patterns, changes in sunlight, and the appearance of new budding in plants. They observe, measure, and report their data via the Web. Scientists use the student data in their own real research and, in return, help students to develop their understanding of the seasons. The analysis of the student data reveals reliable images of animal migration and climate change. Like the JASON Project, Journey North actually predates the Web. Established in 1991 with a grant from the Annenberg Foundation to the Corporation for Public Broadcasting, Journey North was designed to provide innovative science education experiences for K–12 students. In 1995, however, the Web provided the perfect medium for Journey North to achieve its goals, as kids across the country became able to communicate quickly and easily with one another and with the Journey North scientists.

These projects and others that provide access to the resources of science laboratories have helped to humanize the process of science. This is truer of nonprofit governmental science researchers than for-profit labs, but the ultimate result is a culture in which students are empowered to explore and discover the world of science.

SCIENTISTS IN THE LIVING ROOM

The Web puts people—students, teachers, parents, and lifelong learners—closer to the real work of science than they ever have been before. This new reality is probably most impactful on the parent-child relationship.

The average parents expose their children to countless professions in the course of a day. This exposure happens in unintended and unspoken ways, but it happens. Children see bus drivers, teachers, police officers, store clerks, nurses, doctors, bankers, and pharmacists, just to name a few. For most parents, it has never been possible to expose their children to practicing scientists. The Web, however, empowers parents to do just that.

As the Web helps to humanize the process of science, it helps to overturn stereotypes about who can be a scientist. Raise your hand if you remember when your image of a scientist looked a lot like Mr. Wizard. Middle-aged white men with wild hair, dressed in white lab coats, have become a cliché in professional science, but many parents have not yet replaced their mental image of a scientist with something new. The Web offers alternative images as it puts people closer to the real work of professional science than they have ever been.

During the 1960s, television brought science into the living room. Sputnik, President Kennedy, and, yes, Mr. Wizard helped to popularize science at a time when America most needed to inspire a generation of scientists and engineers. Today, television still brings science into the living room; in fact, the number of science-related channels and programming has never been higher. Yet, no comparable generation of scientists is emerging from the audience.

Instead, parents can best use the Web to support their children's science learning by providing opportunities to talk about and wonder about science. The Astronomy Picture of the Day, the Virtual Cave, and the Little Shop of Physics are reliable destinations that provide a quick and easy opportunity for parents to introduce their children to the beauty and wonder of science. Kids need and want to see that their parents value science. When parents make a habit of visiting science-related websites, kids will make a habit of wondering about science.

Since June 1995, the scientists at NASA's Goddard Space Flight Center have selected an Astronomy Picture of the Day. These spectacular images from the cosmos feature simple explanations written by professional astronomers. Like a daily vitamin, a visit to the Astronomy Picture of the Day is a fantastic way to supplement your child's recommended daily allowance of scientific wonderment.

The Virtual Cave is a classic website, first explored in 1995. From the comfort and safety of the living room, parents can take their children on an underground adventure. Solution caves, sea caves, or lava tubes . . . pick your destination for an amazing journey. No matter which you choose, you're sure to see formations that make you say, "Wow!" Solution caves, for example, house the amazing, rare balloon formations. Or look for aragonite formations, which might remind you of snowflakes frozen in rock. So, pop some popcorn, grab your kids, and go caving together in search of cave coral, a common formation, commonly known as . . . popcorn!

The Little Shop of Physics has been an online scientific playground since 1997. The Online Experiments, in particular, are sure to amaze and inspire aspiring scientists of all ages. Some of them involve common household materials, like "Fun with Flatware," "Building Better Bubbles," and "Bernoulli Ball" (which promises "levitation for less than one dollar"!). Perhaps even more interesting, though, are the experiments that you can do using your computer. For example, "Sticky Tape" uses your computer monitor's static electricity to either attract or repel pieces of scotch tape, depending upon the tape's charged particles. "Seeing Spots" uses a drop of water—yes, a small drop of water—on your computer's monitor to demonstrate pixilation. "Head Shrinker" might give you a headache, but it also shows you how optical illusions happen.

These sites and countless others provide rich multimedia that offer powerful visions of science. Animations, videos, and high-resolution imagery invite students to appreciate the beauty and wonder of science. Is it any wonder that kids think textbooks are boring?

A DECADE LATER

It's no coincidence that the projects and sites described above are all marking a decade of service to science education. Good websites endure. It's worth considering, too, that an educational generation has passed from elementary school through middle school and into high school during that same decade.

Today's high school students don't recall a world without the Web. This generation of learners has finely tuned search skills. They scan a list of search results and detect patterns with a critical eye. Digital video, Flash, and web pages are their means of expression. Multitasking is their game. Why do so few of them recognize their readiness for a career in science or engineering?

The "No Child Left Behind" education law has created a classroom environment in which the primary emphasis is on math and reading since the stakes are high for test results in these areas. Nobody would argue the importance of reading and math. However, the unintended consequence is that other subjects—especially science—are getting pushed to the periphery. In many cases, teachers are "sending science home." Is that bad? Not necessarily. When students express their scientific curiosity at home, their parents may begin to recognize and nurture natural talents. Before home computers, parents nurtured scientific curiosity with books and chemistry sets. With the Web, parents can help their children find resources related to any area of science they can imagine.

PROFESSIONAL DEVELOPMENT

For K–8 teachers, the Web has opened the doors to science in unprecedented ways. Most K–8 teachers have little—if any—preparation to teach students about the wonders of science. Many of those teachers don't even appreciate the wonders of science themselves. The Web empowers them to bring science into their lives and into their classrooms in important ways. For teachers, the Web offers opportunities for personal and professional development. Discussion lists, news articles, and webcasts are safe entry points. Collaborative projects and online courses call for more active involvement. These are just a few of the choices on the science smorgasbord that the Web offers teachers. Most teachers help themselves to a balanced serving of risk and reward. The Web becomes most nourishing for teachers, however, when they step into the kitchen and cook for themselves.

Today, most teachers of science in K–12 have developed strategies for using Web resources with their students, especially as more and more of the science content and resources are presented in a way that makes them readily connect to standardized curricula. In fact, many teachers have become quite savvy consumers of science content. Too few, however, have become active producers of Web content. When teachers take hold of the Web as a medium for creative teaching and learning, the impact on students increases exponentially. The Web offers something for everyone. A productive teacher grabs hold and squeezes out something for each one. Teachers can set the scene and shape the moments for each individual student to encounter the wonders of science. Perhaps Maria has previously expressed an interest in star clusters. Jason has a keen interest in black holes. Terry can't stop talking about asteroids. With the Web, their teacher can structure an individualized learning experience for each one of them within the same space science investigation.

The Web also allows teachers to find peers. Online learning circles enable conversation and professional development, helping to overcome the traditional isolation associated with the teaching profession. No longer is a teacher limited to the human and physical resources that exist within the school walls. The Web thrives on open-source software and open content. Teachers thrive on the open classrooms that result.

PRIMARY SOURCES

Traditionally, most science teachers used textbooks as the primary method of classroom instruction. Textbooks are derivative—they derive from primary sources of information, giving students a trickle-down version of reality. The Web brings the primary sources of science into the learning environment.

In the context of science education, what are primary sources? Live weather data. Very large, unfiltered, scientific data sets. Streaming video of live science experiments happening in laboratories. Artifacts that document the history of science.

Science museums—like the Franklin Institute in Philadelphia—hold a treasury of the history of science. Through the Web, learners can explore the artifacts that generations of scientists have left behind. The Franklin Institute Online offers images, animations, and video of selected objects and documents from many of the museum's collections. For example, the Wright Brothers Aeronautical Engineering Collection contains the objects and documents from the 1901 Wright wind tunnel. These artifacts, especially the handcrafted airfoils (miniature wing shapes) that Wilbur and Orville tested in their workshop's wind tunnel, are forensic evidence of how curious young men became pioneers of aviation. Presented online, the collection tells its story to all who pause to wonder; it tells the extraordinary story of two ordinary men who chose to stop and wonder why humans couldn't fly. Their scientific curiosity lifted them to unprecedented heights.

If science museums hold science past, where would you look to find science present? Current science happens every day in laboratories across the country. Through the Web, learners have a window into the production of new scientific knowledge that happens in science laboratories. To find the future of science, of course, you need to look to the next generation: students in classrooms, working with teachers, and children at home, learning with parents. Wherever children feel safe to wonder, the future of science is being written.

ON THE HORIZON

Internet2 is the nation's high-speed research and education network. As a regulated private network, Internet2 is led by over 200 U.S. universities, developing and deploying advanced network applications and technology, and accelerating the creation of tomorrow's Internet in a noncommercial environment. Internet2 is entirely separate from the existing Internet, which is now often called "the commodity Internet." The key benefits of Internet2 are quality and access. Internet2's extremely high bandwidth (at least 10 Mbps) enables high-definition broadcasting, multicasting, videoconferencing, and remote instrumentation.

Increasingly, K–12 schools are establishing connections to Internet2. The Internet2 K–20 initiative has actively promoted usage of the network within the K–12 community. The most recent survey for which official data exist was conducted in May 2004. That survey counted thirty-four statewide K–12/K–20 networks connecting about 23,392 K–12 schools. With nearly 24,000 schools now connected, what are science teachers doing with Internet2?

Teachers in classrooms can, in effect, bring an entire suite of research laboratories into their classroom as they make use of the science research resources that are accessible via Internet2. For example, Lehigh University's Center for Advanced Materials and Nanotechnology has a K–12 science education initiative called "ImagiNations." Teachers and students are invited to operate the lab's scanning electron microscope. With Internet2 bandwidth, the manipulation is real-time. The microscopic world brings the wonder of science into focus. Once you've seen a mosquito's face under the scope, you'll never smack one dead in quite the same way.

On another day, students might "go live" to the Princeton Plasma Physics Laboratory and talk with a plasma physicist as the latest plasma reaction occurs. It happens in an instant, but the plasma energy reaction is captured on high-definition video and—via Internet2—ready for replay immediately. The lab also offers its Internet Plasma Physics Education Experience, which features interactive plasma reaction simulations. Live or simulated, students can participate in plasma energy reactions. Sound like the way you learned science? Probably not.

Internet2 is not ready for the living room yet, but the day when its capacity is accessible from home is not so far in the future. The 100 x 100 Project is piloting the delivery of 100Mbps bandwidth to 100 homes in the Pittsburgh, Pennsylvania, area as a proof-of-concept project. The project's ultimate goal is to deliver 100Mbps to 100 million homes, creating a network that provides universal access with at least the same level of coverage provided by telephone service today. When this kind of network capacity is accessible from home, the opportunities for science education will be limitless.

One thing is for certain: Internet2 is faster, safer, cleaner, and friendlier to the busy classroom teacher than the commodity Internet has ever been. That fact alone suggests that teachers will find innovative ways to use it in their classrooms.

THE BOTTOM LINE

Across the country and around the world, in classrooms and living rooms, kids are using the Web. With a little encouragement and direction, they can use the Web to learn about science. When they use the Web to wonder about science, they discover the whole wide world—which is filled with boundless possibilities.

WEBSITES

Astronomy Picture of the Day: http://antwrp.gsfc.nasa.gov/apod/astropix.html
GLOBE Program: www.globe.gov

100 x 100 Project: http://100x100network.org
Internet2 K–20 Initiative: http://k20.internet2.edu
JASON Project: www.jason.org
Journey North: www.learner.org/jnorth
Lehigh University ImagiNations: www.lehigh.edu/~inimagin
Little Shop of Physics: http://littleshop.physics.colostate.edu/OnlineExperiments/
          OnlineExpts.html
Princeton Plasma Physics Laboratory: http://science-education.pppl.gov
Virtual Cave: www.goodearthgraphics.com/virtcave
Wright Aeronautical Engineering Collection: www.fi.edu/wright

42

Educational Robotics

First State Robotics, Inc.

Education is not the filling of a pail, but the lighting of a fire.
—William Butler Yeats

The four authors who worked on Chapter 42 were: John A. Larock, president of First State Robotics, Inc., who serves as team coordinator for the high school robotics team and works as a human resources manager at DuPont; Carol R. Perrotto, a board member of First State Robotics, Inc., who works as a patent agent at DuPont; Priscilla Zawislak, the coordinator for the Junior FIRST LEGO League Program under First State Robotics, Inc., who works as a quality manager for FMC BioPolymer; and Linda H. Grusenmeyer, the coordinator of the Miracle of Reading Education (MORE) literacy outreach program for First State Robotics, Inc., who works as an educational researcher in the University of Delaware Research and Development Center.

Over the past decade, the field of educational robotics has exploded. More school-aged children are learning about programming, electronics, and problem solving through robots than ever before. Dozens of educational robotics kits and curricula spread across the country in a myriad of forms. Robots, once only fiction, have become part of our daily lives. Hundreds of thousands of families are cleaning their homes with robot vacuums. Two twin robotic rovers have been exploring the surface of Mars for more than a year. Worldwide investment in industrial robots is up 19 percent in 2003, with worldwide growth predicted at an annual rate of 7 percent in the period

2004–2007. More robots are seen in roles of service to society as part of bomb squads, surveillance organizations, military operations, and assistive technology for people with limited abilities.

Educators, organizations, and parents alike are utilizing the wonder and excitement of educational robotics to "light a fire" in today's youth to learn about technology and to actually participate in its creation.

DEFINITIONS AND HISTORY

The term *robot* was first used by Czech playwright Karel Capek in his play *R.U.R. (Rossum's Universal Robots)* in 1921. Robot is derived from the Czech word *robota*, meaning drudgery or slavelike labor. Robots became possible in the 1950s and 1960s due to the enabling technologies of transistors and integrated circuits. There are numerous definitions of robots, but most agree that a robot is a programmable machine that imitates the actions or appearance of an intelligent creature—usually a human. The machine must meet two requirements to be a robot: (1) obtains information from its surroundings, and (2) does something physical, such as move or manipulate objects, based on that information.

Robotics as a field of technological study began in the 1970s. Although educational robotics has existed in some form since this time, it has been within the past ten years that the field has exploded. Each month, numerous articles on robots or robotics grace a number of national publications. Educational robots can be perceived as the use of robots and robotics to educate individuals or groups in areas of science, engineering, and technology.

SCIENCE EDUCATION THROUGH ROBOTICS

With technological advancements of robot microcontrollers, like the Handy Board, developed by MIT, robotics projects and kits are more widely available to hobbyists, educators, and industry. This availability has contributed to the growth of robotics education, both formal and informal.

Skills acquired from robotics learning include hard technical skills and soft project skills, both critical for successful engineers and scientists.

"Hard Skills"

Computer programming

Autonomous robot control and operation

Electronics and electrical engineering

Mechanical engineering

Drafting and design

Systems engineering

"Soft Skills"

Creative problem solving

Teamwork and collaboration

Communication skills

Confidence and responsibility

Project planning

Adaptability

THE MOTIVATION FACTOR OF ROBOTICS

Many studies have shown that when students of any age group are participating in an area they enjoy, they spend more time working in that area and are
more likely to learn. Motivation is in large part governed by interest. Robotics
has the unique benefit of bridging the gap between today and the future.
Children see real robots exploring Mars and also can feel comfortable learning
to build their own robots with LEGO Mindstorms. Almost immediately, they
can make a connection between their work and opportunities of the future.
Recently, the FIRST LEGO League Challenge, "No Limits," required students
to design robots to assist in accomplishing common tasks which people with
physical limitations may not be able to do. Some of the design ideas developed
by middle school students may result in actual patents.

Attempting to navigate the current world of educational robotics is challenging. Programs are divided by age group and by various focus areas. This
chapter provides a roadmap for those interested in educational robotics by
looking at existing programs in the areas of robotics curriculum, robotics kits/
toys, and robot contests.

ROBOTICS CURRICULA

Many sources of robotic curricula exist online; some are based around
individual competitions, and others more general. Several sites have compiled
a large number in one location for easy access. Not all curricula available
online are free, however, and registration may be required as well.

The NASA's Robotics Curriculum Clearinghouse (NASA 2006) is an online collection of featured robotics curricula ranging from full courses to
single-lesson plans that can be readily incorporated into classroom activities.
It is searchable by subject, grade level, state standard, language, teaching
method, cost of materials, and many other fields. The site also includes a
clearinghouse of K–12 robotics-related educational materials, posted by users
of the site (NASA 2005).

The Carnegie Mellon University and the National Robotics Engineering
Consortium Robot Academy (Carnegie Mellon University n.d.) includes many
resources, including robotics camps, clubs, competitions, internships, K–12
education, teacher training, curriculum development, and multimedia

support. The curricula are geared toward K–12 students and designed to be used with the LEGO Team Challenge Mindstorm Robotics Set and with ROBOLAB software. Free lesson plans are included, but the full curricula must be purchased; they are also available in Spanish. Links are available for many other sources of curricula.

The Educational Robotics Cyber Laboratory (n.d.) is a curriculum site for teachers, students, and other users who are interested in educational robotics. Many of the laboratory activities are especially designed for participants of the KISS Institute for Practical Robotics Botball program; however, they can be used by anyone with a LEGO Mindstorm kit, a LEGO Technic kit, or a Handy Board processor. More advanced activities and research projects are provided to challenge experienced robot designers.

BattleBots IQ (n.d.) is an educational program created by the producers of the *BattleBots* television series. It can be used as a stand-alone robotics class, or teachers can integrate the curriculum into an existing class or run the BattleBots IQ program as an after-school club. BattleBots IQ National Competition is an event held every spring as the culmination of the BattleBots IQ Robotic Curriculum. At the competition, students from middle schools, high schools, colleges, universities, and technology centers converge to test their final robotic creations. Training is available at the BattleBots IQ Teacher Training Institute.

The Ottawa-Carleton Catholic School Board has a program for grades 4–6 called Robotics in the Classroom (n.d.). They have created a mobile classroom called RoboDome that offers elementary students a chance to benefit from being immersed in a hands-on learning environment where they will design, build, and program robots. It is based on Ontario's Math and Science expectations for the junior level.

The Massachusetts Science and Technology/Engineering Curriculum Frameworks Center for Engineering Educational Outreach (2001–2003) integrates engineering concepts and activities into pre-K through twelfth grade classrooms. The activities available on the site are linked to the standards for technology literacy (International Technology Education Association). It includes activities for all grade levels and many resource links.

ROBOTICS TOYS AND KITS

Robot resources are now available for children from three years old through high school and beyond. For the younger age group, numerous popular robot toys are available.

The Kinderbot (Fisher-Price) is aimed at children from three to seven. The Kinderbot contains a large variety of learning and interactive tools, including games, singing, and dancing. Versions for even younger children (Toby the Totbot, for ages one and up; and Fetch the Phonicsbot, for ages two and up) are also available.

Robotics imitating human beings, as we are used to seeing in film and television, are beginning to enter our real world. Asimo, a research project from Honda, is a life-size robot that walks. The Robosapien (Wow Wee) was designed by a NASA scientist and has literally thousands of moves in his robotic library, including walking, picking up objects, throwing objects, and even break-dancing.

These toys provide an environment where youth can be comfortable interacting with robots. This environment can encourage children, who are do-it-more-yourself robot enthusiasts, to examine robot kits as they get older.

Robot kits and parts can be purchased at many sources. Some have pre-programmed and partially assembled kits; others are basic parts used for your own designs. Table 42.1 contains information on a number of robot kits. Some include school curricula. This list is by no means all-inclusive, but provides a starting point for parents and educators to explore their options.

ROBOT TOURNAMENTS AND CONTESTS

[FIRST Robotics Competition] gives kids the chance to fail—an important creative opportunity. Creativity is messy, it's risky, it makes your brain hurt, and sometimes people don't believe you. But creativity and a robust self-image are fundamental life skills. If we reward creativity at this stage (high school), then they will try it when the stakes are higher.
—Woodie Flowers, Pappalardo Professor of Mechanical Engineering, Massachusetts Institute of Technology

One of the most fun and successful means to develop students' interests and understandings of science and technology is through participation in robotics contests. These provide an opportunity for all students to learn how science and technology are used to solve real-world problems and that problem solving in a team can be fun. Contests provide hands-on experience in applying math, science, technology, engineering, and computer concepts while developing students' teamwork and leadership skills. Students learn that each team member can contribute to making the project better by participating through brainstorming, research, planning, exchanging ideas, testing, building, and competing or by making presentations. Survey results from these programs have demonstrated that these provide a positive experience for both boys and girls. They learn that the science and computer skills used in these contests are not much different than those needed for careers in these fields. In addition, this is often their first exposure to teamwork, and this, perhaps, is the most significant impact of robotics programs for all students, regardless of their interest in any particular subject. The experience of being an important part of a team, having responsibilities, having adults pay attention to them, and making important decisions gives them a high degree of satisfaction and grows their self-confidence.

Contests have a significant impact on teachers as well. Many report that their own use and understanding of science and technology in the classroom

TABLE 42.1 Basic Information on Robot Kits

Kit Name	Company	Features	Educational Target	Contact Information
Botball Kit	KISS Institute	Fully autonomous mobile robot	Middle and high school	www.kipr.org/products/index.html
BYO-Bot	KISS Institute	Pre-assembled. Not programmed directly by student.	Middle and high school	www.kipr.org/products/index.html
Eggo Robotics Platform	Eggo Educational Robotics	Compatible with LEGO Robotics using uCLinux	Middle and high school	http://eggo.sourceforge.net/
ER1 Robot Kit	Evolution Robotics	Step-by-step exercises to get students started in robotics	Middle school, high school, and college	www.evolution.com/education/
GEARS-IDS Invention and Design System	GEARS Educational Systems LLC	Kits and accessories for use in the classroom	Middle and high school	www.gearseds.com
Intelli-Brain Bot	RidgeSoft LLC	Java-enabled software	Middle and high school	www.bestinc.org
K'NEX Education	K'NEX		Elementary and middle school	www.kbotworld.com
LEGO Mindstorms	LEGO		Elementary and middle school	www.lego.com/eng/products/next/
Robodyssey	Robodyssey Systems, LLC	Robotics platform outgrowth from New Jersey education project	Middle and high school	www.robodyssey.com
Vex Robotics	Innovation First, Inc.	Autonomous or driver-controlled mobile robot with variety of sensors	Middle and high school	www.vexlabs.com
World of Robotics Online	World of Robotics	Key learning areas across grade levels	Elementary, middle, and high school	www.robotics.com.au

are enhanced, and they become more effective teachers. Classroom learning is reinforced while encouraging abstract thought, creativity, problem solving, and self-directed learning. They realize that students can accomplish much more than they originally expected. Contests also provide teachers with a means to promote good sportsmanship, leadership, and school spirit. Some contests have workshops for teachers where they are coached in how to design, build, and program a robot. These workshops also help teachers learn about current robotics technology and how to integrate it with curriculum. Alternatively, many parents, colleges, and corporations provide professionals willing to help teachers as mentors and provide the technical skills needed.

Increasing the diversity among participants is a focus of many robotics organizations. Many robotics programs promote forming partnerships with organizations to collaborate with educators in providing role models and encouragement for building the future workforce of scientists, technologists, and engineers. These can have a significant impact in sparking interest and providing opportunities for students not typically exposed to these career possibilities.

A variety of contests are available for all age groups and abilities. Some of the larger contests, such as FIRST and RoboCup, have programs that build on similar themes at each level and provide a means for students to continue with new challenges as they progress from elementary to high school. Contests range from international to local competitions and vary in length from a few weeks to months. Costs can vary from free to several thousand dollars. Contests can have a consistent format, which uses the same challenge as in previous years, or they may present a new challenge each time. Some contests are performance based; others have robots that compete against one another. Many of the robotics contests use LEGO Mindstorms kits, RoboLab kits, or their own kits for robot construction. Others are open-ended, allowing a wide variety of materials and components. Table 42.2 summarizes some of the more widely known tournaments and contests.

GETTING INVOLVED

Educational robotics is evolving, and it is exciting groups of all ages about science and technology. Opportunities to participate in this evolution are ample. Educators now have multiple curricula and associated kits and resources to incorporate robotics in the classroom. Parents can enjoy the excitement themselves as they connect children to a number of robotic contests or off-the-shelf kits. Students have many options to learn about robotics, from programs starting in elementary school all the way through college.

By participating in educational robotics programs, individuals can contribute to "lighting the fire" for technology interests in our youth that, hopefully, will never be extinguished.

TABLE 42.2 Examples of Robot Tournaments and Competitions

Contest	Organization	Format	Educational Target	Contact Information
FIRST Robotics Competition	FIRST	New Challenge	High school	www.usfirst.org
FIRST LEGO League	FIRST	New Challenge	Elementary and middle school	www.usfirst.org
Jr. FIRST LEGO League	FIRST	New Challenge	Elementary school	www.usfirst.org
BEST	BEST	New Challenge	Middle and high school	www.bestinc.org
Botball	KISS Institute for Practical Robotics	New Challenge	Middle and high school	www.botball.org; www.kipr.org
LEGO Simple Machines	NASA	New Challenge	Elementary school	
NASA Robotics Education Project	NASA	Sumo, Line Running, and Line Following Challenges	Elementary, middle, and high school	www.robotics.nasa.gov students.challenge.htm
RoboCup	International	Soccer Leagues for Robots	High school	www.robocup.cup.com
RoboCup Jr.	International	Dance, Rescue, and Soccer Challenges	Elementary, middle, and high school	www.robocupjr.com
Robofest	Lawrence Technological University	Robotics Missions Emulate Real-World Environment	Elementary, middle, and high school	www.robofest.net
Trinity College Fire Fighting Robot Contest	Trinity College	Firefighting in a model house	Elementary, middle, and high school	www.trincoll.edu/ events/robot/
Penn State Abington Mobile Robotics Program	Penn State University, Abington, Pennsylvania	Robo-Hoops, Firefighting, Outdoor Robot Contest	Elementary, middle, and high school	www.ecsel.psu.edu

ABLE 42.2 *continued*

ontest	Organization	Format	Educational Target	Contact Information
attleBots IQ	BattleBots	New Challenge	High school	www.battlebots.com
'NEX K*Bot 'orld hampi- nship	K'NEX	New Challenge	Elementary, middle, and high school	www.kbotworld.com

REFERENCES

Battlebots IQ. N.d. Battlebots IQ: The Smart Sport. www.battlebotsiq.com.
Carnegie Mellon University, National Robotics Engineering Center. N.d. Robotics Academy. www.rec.ri.cmu.edu/education/.
KISS Institute for Practical Robotics. N.d. Content Learning Activities: Educational Robotics Cyber Laboratory. www.kipr.org/curriculum/content.html.
NASA. 2005. Robotics Education Project. http://robotics.arc.nasa.gov/a_educators .htm.
———. 2006. Robotics Curriculum Clearinghouse. http://robotics.nasa.gov/rcc/.
PreK-12 Engineering. 2001–2003. Welcome. www.prek-12engineering.org.
Robotics in the Classroom. N.d. www.occdsb.on.ca/%7Eproj4632/.

ADDITIONAL RESOURCES

Center for Youth and Communities. Evaluation of the FIRST LEGO League. New York: Center for Youth and Communities, Brandeis University.
Meeden, Lisa, Alan Schultz, Tucker Balch, Rahul Bhargava, Karen Zita Haigh, Marc Böhlen, Cathryne Stein, and David Miller. 2000. The AAAI 1999 Mobile Robot Competitions and Exhibition. American Association for Artificial Intelligence. *AI Magazine*, Fall, 69–78.
Murphy, Patricia J. 1997. National Robotics Competition Brings out the Fun in Learning. *The Technology Teacher* 66 (October): 29–31.
Nourbakhsh, Illah, David Miller, Corinna Lathan, and Maja Mataric. 2004. *Educational Robotics: Assessment of the State of the Art in the US*. July. Washington, DC: National Science Foundation.
Tesler, Pearl. 2000. Universal Robots: The History and Workings of Robotics. www.thetech.org/exhibits/online/robotics/universal/index.html.
United Nations Economic Commission for Europe. 2004. *World Robotics Survey*. October 20. www.unece.org/press/pr2004/04robots_index.htm.
Verner, Igor M., and Eyal Hershko. 2003. School Graduation Project in Robot Design: A Case Study of Team Learning Experiences and Outcomes. *Journal of Technology Education* 14 (Spring): 40–55.

43

Science Teaching and Learning through the Zoo, Aquarium, and Botanical Garden

Christine Klein

A growing number of zoos, aquariums, botanical gardens, arboretums, science centers, planetariums, museums, nature centers, and other science-rich institutions offer programs and resources to teachers, students, administrators, preservice teachers, teacher educators, and parents. These institutions, sometimes called *informal science education institutions*, are not just for field trips or tourists. They offer a variety of innovative materials and ideas for students, families, adults, and educators. This chapter looks at how educators can tap into and maximize the benefit they get from the resources of zoos, aquariums, and botanical gardens. Much of the information in this chapter applies to families and individuals outside of the school system, though the focus is on how teachers and schools can best use these resources.

To understand the unique opportunities available, I begin with an exploration of the nature of learning in these institutions. Next, I discuss the traditional approaches and institutions use to support schools. More recent developments in educational programming, albeit atypical, are then examined. In the final section, I suggest ways to take advantage of these resources by offering a sample lesson plan.

SCIENCE EDUCATION IN INFORMAL SCIENCE INSTITUTIONS

What is *informal science education*? Many authors have written articles discussing the terms *informal* and *formal* as applied to education and educational institutions. In the 1970s, museums, zoos, and similar institutions wanted to identify themselves as places of learning while distinguishing themselves from schools and the type of learning that occurred in schools. If schools were *formal institutions* where learning followed *formal* structures, then the science institutions could call themselves *informal science education institutions* where *informal learning* occurred. These terms persist today in science education literature. Exploring them helps us to better understand the role of these institutions in science education as we look at what makes each setting unique in terms of learning opportunities.

LEARNING IN INFORMAL SETTINGS

The term *informal* is most often defined as "that which is not formal." By defining something by what it is not, we leave the category wide open. Does informal learning include all learning outside of school, even watching the Discovery Channel from my couch? Most authors who use *informal science education* or *informal learning* refer to learning that occurs outside of schools, such as in zoos, after-school programs, and libraries. Such learning occurs throughout a person's life; is self-directed, voluntary, and not part of school curriculum; and often occurs in a social context.

Many people interested in the learning in zoos, aquariums, and botanical gardens have noted similar characteristics of learning there and in schools. We can hear a formal lecture at a zoo or engage in an informal self-directed inquiry in a school classroom. Along with others, I have argued that we should not use the terms *informal* and *formal* to describe learning. John Falk and Lynn Dierking suggest the term *free-choice learning* as an alternative, which they define as "free-choice, nonsequential, self-paced, and voluntary" and "primarily driven by the unique intrinsic needs and interests of the learner" (Falk 2001, 7). Some educational staff members of zoos, aquariums, and botanical gardens recognize that their own programs developed to support school curricula do not meet the definition of free-choice learning, and so continue to look for new language.

Regardless of the terms, instructional strategies that typically occur in the classroom are different from those employed by family visitors to an aquarium or similar institution. In workshops with preservice and in-service teachers on utilizing resources of the St. Louis institutions, I ask them to spend an hour observing a family group and a school group in an exhibit area, noting the strategies used by each group. Most teachers note differences in the adult-child interactions, including who decides where to go and to what to attend. They note the free-choice nature of the family groups as they follow their own interests and spend time discussing and trying exhibits. They note the structured nature of the school group, which all too often involves a scavenger hunt and children moving quickly from one exhibit to another.

What are effective instructional strategies to maximize the resources of these institutions? We shall see below.

UNIQUE FEATURES OF ZOOS, AQUARIUMS, AND BOTANICAL GARDENS

Zoos, aquariums, and botanical gardens are unique among informal science learning environments with their live plants and animals and with their conservation and biological research. At most of these institutions, practicing scientists and educators work side by side to study and preserve biodiversity and to communicate the importance of caring for our earth.

Most zoos, aquariums, and botanical gardens have live animals and plants that children can touch or hold. To understand "adaptation," a child can read about it in a book, create examples of plant and animal adaptations through educational activities, see live plants and animals at an informal science institution and observe their adaptive characteristics, or learn about adaptations firsthand in natural settings. Utilizing all these strategies in a unit of study can make the learning richer and perhaps last longer. When face-to-face with a naked mole rat at a zoo, a child appreciates its adaptations for living underground in hard, dry clay. Seeing its large teeth, strong jaw muscles, and burrowing behaviors firsthand creates a lasting memory of animal adaptation. Subsequently, reading about naked mole rats in a children's magazine, or reading about other animal adaptations in the library and creating hypothetical plants or animals adapted to an environment, combines with the zoo visit to extend and deepen the learning.

How can we maximize learning in school settings with the unique resources of zoos, aquariums, and botanical gardens? We need to move beyond the traditional understanding of field trips, but we can start there and expand.

TRADITIONAL APPROACHES TO EDUCATIONAL PROGRAMS

Informal science education institutions have been the sites of field trips for school groups since the nineteenth century. However, a look at the list of the educational offerings of institutions reveals many and more diverse opportunities. Teachers can book structured programs led by staff or volunteers in conjunction with their field trip, book outreach programs in which institution staff come to their school, or borrow resources available through loan programs.

Many teachers say they don't know what resources are available in their area. In this section, I provide examples of some of the more traditional opportunities.

Field Trips and Structured Programs

Perhaps you remember a field trip to a zoo as a student. Was the trip part of your schoolwork or a reward for something your class did well? Do you remember what animals you saw or what you learned? Perhaps you remember a field trip as a teacher or parent chaperone. Did you have to make all the connections to the curriculum yourself, or was there a zoo program to enrich your curriculum? Was zoo staff available to provide information on logistics or on the science content?

Most zoos, aquariums, and botanical gardens have moved beyond the practice of opening their doors to school groups, simply saying, "Y'all come." To help teachers make the most of their visit on a field trip, most institutions offer field trip planning online, by phone, or through mailings. Some institutions offer curriculum-based activities for use in the classroom before and after the visit.

During the field trip, many institutions allow school groups to visit much as a family group would, with the freedom to choose where and when they spend their time. Many also provide guided tours to groups, providing students the chance to see plants and animals with someone to answer their questions and stimulate their interest. By enrolling a class in a structured program designed to meet state or national standards, there is even more opportunity to direct students' attention toward specific learning goals.

Research on successful group visits to museums, zoos, and other informal science education institutions suggests a number of effective strategies (summarized by Janette Griffin; see Griffin 2004). Careful planning on the part of the educator is important. Intentional links to the curriculum, personal knowledge of the institution (e.g., knowing what is open and what is available on the day of the visit), and connections to the lives of the students in previsit, visit, and postvisit activities improve the potential for long-term interest and learning. Varying the activities during the visit, increasing firsthand experiences with live plants and animals, and decreasing the use of worksheets can also improve the effectiveness of the visit. Students who have been oriented and know what to expect during the visit are better able to focus on learning. For example, the South Carolina Aquarium offers an orientation video for students.

Outreach Programs

As funds for buses diminish in school budgets, institutions are finding ways to continue to meet school needs. Many are taking their programs, plants, and animals on the road. Staff in vans can usually travel anywhere within a day's drive. For example, the Audubon Nature Institute in New Orleans delivers programs to students within a fifty-mile radius through their BugMobile, AquaVan, or ZooMobile. Programs at some institutions bring educational materials and real artifacts (bones, animal coverings, shells, etc.) to the classroom without the live plants or animals. As with the field trip, to maximize the learning potential of the outreach program, links are needed to the ongoing class curriculum through pre- and postprogram activities, which are often provided by the institution.

Kits and Trunks

Educators who prefer to use resources on their own may borrow kits or trunks from an institution. For example, the Indianapolis Zoo offers the Project

Elephant Kit; Feathers, Fins and Furry Tales Trunk; Suitcase for Survival; and Snake Kit, each with curriculum and materials. Such resources provide great potential for explicit curriculum links. At the St. Louis Zoo, kits are often developed around particular topics because of specific teacher requests.

INNOVATIVE APPROACHES TO EDUCATIONAL PROGRAMS

Informal science education institutions have begun to do much more to support schools and science curriculum over the last twenty-five years. Most start with field trips and structured programs at their facility. As they gain experience, many offer a wider variety of programs to schools and community groups at the school or other sites (Inverness Research Associates 1996). With increased experience and with larger educational staffs, many institutions are beginning to offer more sustained programs or more technologically advanced programs.

Overnight Programs

Some zoos provide overnight programs so that children and adults can see nocturnal animals engaged in their natural behaviors. Educational activities, games, and crafts often augment a night hike. Even with this type of memorable experience, careful planning connecting the activity to curriculum and orienting students can improve the learning experience.

Distance Learning

At the Indianapolis Zoo, a teacher may enroll his or her class in a scheduled distance learning program or even create a specialized program in conjunction with zoo staff. During the broadcast, a zoo staff member provides live footage of animals in their naturalistic exhibit-based habitats with the opportunity for students to ask zoo staff questions in real time. Video clips, artifacts, slides, and activities are used to create a full interactive experience. With such distance learning opportunities growing in popularity, students without easy access to a science-rich institution can still take advantage of their resources.

Webcams, Streamed Videos, and Other Online Resources

With Internet access, you can explore the various webcams now focused on animals in zoos and aquariums around the world. These cameras capture animals in their naturalistic exhibit areas as they carry out their daily activities. Elephants, polar bears, pandas, and apes appear on the San Diego Zoo's website; belugas at the Vancouver Aquarium; and penguins at the Montréal Biodôme, among a growing list animals and websites. Not all animals are within view of the camera at all times, not all animals are awake or active when students are watching, and not all webcams are functional all day, but peeking in on an animal at your local zoo or at an aquarium around the world can pique student interest as you begin a unit on one particular animal, a specific habitat, or a general biological concept.

If webcams don't meet your needs, some websites provide short, streamed video clips. At the Honolulu Zoo's website, you can view a variety of animals in their enclosures or at the vet. More traditional video formats are also available for purchase from institutions, like those of the Missouri Botanical Garden focused on different world biomes and various aspects of plant life, and videos from the Brookfield Zoo of animals in various habitats.

Many websites have plant and animal information, photos, and additional information. From the Missouri Botanical Garden's website, you can access information on plants from a variety of databases. For a more kid-friendly website, the American Zoo and Aquarium Association's (AZA) children's website, Aza's Web, provides information on "Cool Critters" with facts and stories. The Houston Zoo's website provides guidance to older students on how to research an animal, with numerous links to other resources. For teachers, many websites have lesson plans and curricula. The National Aviary in Pittsburgh's unique web curriculum can be accessed by students around the world, with online student activities and a linked teacher's guide full of detailed background information.

PRESERVICE TEACHER PROGRAMS
AND PROFESSIONAL DEVELOPMENT

In 2002, the Houston Zoo received the Significant Achievement Award from the American Zoo and Aquarium Association for its "Project U.S.A.: Urban Scientists in Action." Preservice teachers at Texas Southern University worked with zoo staff to develop six-week curriculum units to complement their children's zoo. Six teachers from Houston Independent School District piloted the units with the help of local high school interns. Internships for preservice and in-service teachers often involve curriculum development as a learning component, along with delivery of programs.

The St. Louis Zoo invites professors and their science methods classes to the zoo to learn more about this community resource. Many zoos, aquariums, and botanical gardens offer professional development to teachers, often for graduate credit. Opportunities range from short (two-hour) workshops to extended courses that span a year or more. Some programs, like the Natural Science Institute at the Missouri Botanical Garden, offer yearlong professional development to teachers with rich, live resources and support in using those resources in the classroom.

Many institutions create newsletters specifically for teachers and offer free previews of new exhibits and programs. Some have teacher advisory boards to help guide their education departments. Teacher resource centers are available at some institutions, providing resources that allow educators the opportunity to borrow, copy, or create their own classroom materials, like video clips of selected animals.

PROGRAMS FOR TEENS

A small but growing number of zoos, aquariums, and botanical gardens are joining their museum colleagues in offering programs for teens, a trend that started with the YouthALIVE! program through the Association of Science-Technology Centers. In these programs, youth engage in learning, volunteer, and work opportunities. For example, at the Audubon Zoo in New Orleans, youth work as volunteers in the Junior Keeper (seventh and eighth grades) and the Zoo Corp (high school) programs. At the National Aquarium in Baltimore, high school youth participate in Aquarium on Wheels to learn marine science and develop work and presentation skills.

In programs for youth who will be the first to attend college in their family, college application and preparation are key program elements. The Chicago Botanic Garden offers mentoring and paid internships for high school youth in its College First program and field-based experiences for middle schoolers to gain comfort, interest, and understanding of science in its Science First and Primero la Ciencia programs, the latter a program for Latino youth.

CREATING YOUR OWN LEARNING EXPERIENCES

Now that we know what resources are out there, where do we start in planning for *our* school? How can we integrate these resources into *our* curriculum? In this section, I outline a unit plan, describing steps to design the unit and gather resources.

The following sample middle school unit is based on integrated curriculum units developed through the Schools for Thought (SFT) Collaborative, funded by the James S. McDonnell Foundation in the 1990s. Some SFT units are still implemented at Compton-Drew Investigative Learning Center Middle School at the St. Louis Science Center (Klein 1998), although a version of this zoo unit was conceived with teachers in the pilot phase and is not in use. The plan below does not assume a team effort, though a team approach would be ideal.

Planning the Unit

We begin planning with the key concepts to address, referring to state standards. In science we plan to address the fundamental unity underlying the diversity of all living things, the interdependence of all living things, and the interactions of living things with their environments. If including math, we would add measurement and geometry; if social studies, geography; and if communication arts, persuasive writing and research skills.

After brainstorming ways to motivate student learning and thinking through project ideas, we decide to have students design a zoo as their final project. Planning backward from the final assessment, we outline the criteria for the project as a foundation for the rubric to assess students' understanding of the concepts.

- Six exhibit areas representing six different biomes.
- Each exhibit area has animals and plants from one specific geographical region.
- Each group submits scaled drawings of their zoo with labels for each exhibit area indicating the biome and geographical region.
- Each group submits a report identifying the plants and animals in each exhibit area with rationale for why those particular plants and animals were selected.

Finding the Resources

We know our budget can support a field trip but little else. Our first stop in finding resources is the Internet. We find helpful websites for students and create a list of links to distribute. At each zoo, aquarium, and botanic garden website, we look for teacher resources and a map of the institution. We know we want students to see a variety of maps. We find that some zoos arrange exhibits by habitat like the Audubon Zoo in New Orleans with its swamp, Australian outback, and jungle. Others arrange exhibits by type of animal like the San Diego Zoo's areas for hoofs and horns, cats, and bears.

We look for videos and photos of zoos. None of the videos seem right. We need a hook to start the unit and something to orient students for the field trip. We decide to make our own video of our own zoo. Using a borrowed video camera with a colleague, we make a silent film of different exhibits, carefully selecting from different biomes. To make it interesting, we film each other looking at some exhibits to see if students are paying attention. While at the zoo, we pick up enough maps for each group of students.

Timeline

The timeline below provides an overview of the unit and depends largely on the amount of science class time in a given week.

Week 1
- Introductory video, class discussion
- Introducing the challenge and generating questions
- Biome groups identified

Week 2
- Biome groups conduct research
- Whole class activity: Internet research
- Biome groups report to whole class

Week 3

- Zoo map activity
- Zoo design discussion, zoo groups identified
- Field trip orientation
- Zoo groups design datasheets

Week 4

- Zoo field trip
- Debrief trip with whole class
- Design groups conduct research

Week 5

- Design groups give updates to whole class
- Design groups continue research

Week 6

- Designs finalized
- Presentations by design groups
- Scoring of projects (self- and peer-assessment)

Telling the Story

My favorite step in curriculum development is "telling the story," a process I developed at Compton-Drew. If we told a story about this unit once it was over, what would it say about the experiences of the students? Thinking through such a story *before* the unit is implemented helps identify details that might otherwise get overlooked. This is particularly useful in working with a team of teachers, allowing everyone to ask questions of the story to check their own understanding of the sequence and desired outcomes of the unit. Even though storytelling comes early in the planning process, it is fitting to end this chapter with this unit's story.

As students arrive in their science class, they notice the VCR set up in the front, posters of different natural areas covering one wall, and "Welcome Designers!" written on the chalk board. Ms. Ingals starts the video, a brief look at scenes from the local zoo. A few students recognize the zoo. Then students notice Ms. Ingals waving from behind an exhibit of grassland animals. All students become more attentive, wondering where else she might appear.

After the short, silent video, Ms. Ingals leads a discussion guiding students to note the different types of plants and animals located in different exhibits. "What animals did you notice?" and "Why do you think they put those two animals together?" she asks in reference to student observations. "Why were those particular plants there?" "Why was there a stream in that exhibit area?"

The next day Ms. Ingals introduces the challenge. Design a zoo with six exhibit areas, each from a different geographical region representing a different biome. She reviews the rubric for scoring student projects, then leads students in generating questions. "What questions do we need to answer before we design our zoos?" All students have a chance to speak, and Ms. Ingals guides students toward grouping questions. Some focus on the biomes: "What plants and animals go with each biome?" Some questions focus on zoos: "How are zoos designed?" and "What are the parts of a zoo?"

Ms. Ingals explains how the groups will work for the projects, the jigsaw technique. For the first part of the unit, the class will be divided into six *biome groups* (deciduous forest, desert, rainforest, grassland, tundra, and taiga). Each student in the biome group will need to become an expert in his or her biome because in the second half of the unit, each student will represent a biome in a new *design group*. Therefore, each design group will have one (or two) students with expertise in each of the six biomes and will actually design the zoo.

After a short lesson on biomes and review of the student-generated questions that relate to biomes, Ms. Ingals guides students to recognize that they need to identify a variety of animals and plants for each biome along with each animal's food, water source, and home. They agree on criteria for the biome groups that will help them design their zoos later.

Groups begin their research and develop a sense of the task before them. Ms. Ingals conducts a whole-class review on Internet research, introducing the list of helpful websites. Later, Ms. Ingals spends time with each group to monitor their progress. As groups near research completion, they present their information to the rest of the class for feedback. Then they complete their work, incorporating feedback from peers.

With expertise on biomes, students are ready to design zoos. First, Ms. Ingals brings out maps of several zoos for a whole-class activity to identify the important features of zoos, examine the various ways zoos are organized, and discuss the scaled drawings needed for their projects.

Finally, students are assigned to a design group with representatives from each of the biomes, and they start their designs. Later the same week, students are oriented to their local zoo, using the zoo's map and teacher-made video. Groups create their own data collection sheets for the trip to the zoo. They map out their path and plan to collect information. Ms. Ingals works with groups individually to ensure that students plan to collect appropriate information.

On field trip day, students are excited and eager to collect information. Ms. Ingals has copies of each group's data sheets and map for herself and the chaperones. With one chaperone per group, groups are allowed to follow the paths and plans they developed. Ms. Ingals doesn't assign herself a group. Instead, she checks in with each group at a cluster of exhibits they visit. She notices groups are collecting good information and are taking time to watch the animals. The groups arrive at the predetermined meeting place about forty minutes before the bus arrives. This gives Ms. Ingals time to sit with students to review their information and observations before they leave the zoo setting. She leads a discussion by asking, "What biomes did you see?" and "How will you use the information you collected to design your own zoo?"

At school the next day, design groups work on their projects. Ms. Ingals spends time with each group, redirecting or suggesting resources. She notes that many of the same animals appear in all the zoo designs, but each design is unique. As she monitors progress, she occasionally suggests students return to zoo, aquarium, and botanical garden websites and library reference books to check details. As groups near design completion, they present their designs to the whole class. Peers give feedback, asking about missing or confusing information.

During the last week, each group finalizes its zoo. Groups use the rubric to score their own projects and those of their peers. Ms. Ingals scores projects too. On Friday, parents, students from younger grades, and invited zoo staff arrive for their presentations. Ms. Ingals surprises the class with a "welcoming speech" to the visitors. As groups present their zoos to guests walking from group to group, guests ask questions and are impressed with the students' detailed responses. The unit is a success. Student presentations demonstrate to Ms. Ingals and the guests that the students understand the fundamental unity underlying the diversity of all living things, the interdependence of all living things, and the interactions of living things with their environments.

ACKNOWLEDGMENT

Sharon Kassing at the St. Louis Zoo provided feedback from the zoo perspective, while Jennifer McLelland, assistant principal, and teachers at Tahoka Elementary School provided feedback from a rural perspective. My thanks to all.

REFERENCES

Falk, John H., ed. 2001. *Free-Choice Science Education.* New York: Teachers College Press.

Griffin, Janette. 2004. Research on Students and Museums: Looking More Closely at the Students in School Groups. *Science Education* 88 (suppl.): S59–S70.

Inverness Research Associates. 1996. *An Invisible Infrastructure: Institutions of Informal Science Education.* Washington, DC: Association of Science-Technology Centers.

Klein, Christine. 1998. Putting Theory into Practice: Compton-Drew Investigative Learning Center. *Journal of Museum Education* 3 (2): 8–10.

ADDITIONAL RESOURCES

Suggested Readings

Curator: A Museum Journal. 2004. Special issue devoted to zoos and aquariums. 47 (3): 233–361.

Middlebrooks, Sally. 1999. *Preparing Tomorrow's Teachers: Preservice Partnerships between Science Museums and Colleges.* Washington, DC: Association of Science-Technology Centers.

Websites

Associations with Links to Local Institutions and Educational Resources

American Association of Botanical Gardens and Arboreta: http://aabga.org
American Zoo and Aquariums Association: http://aza.org
Association of Science-Technology Centers: http://astc.org
Botanic Gardens Conservation International: www.bgci.org.uk
Canadian Association of Zoos & Aquariums: www.caza.ca
World Association of Zoos and Aquariums: www.waza.org

Websites for Kids with Parent and Teacher Resources

Aza's Web: www.azaweb.com
Botanic Gardens Trust in Sydney (stories and songs): www.rbgsyd.nsw.gov.au
San Diego Zoo (crafts, recipes, and more): www.sandiegozoo.org

Free-Choice and Museum Learning, Including Bibliographies

Informal Science website at the University of Pittsburgh: www.informalscience.org
Information and videos on terrestrial biomes and aquatic ecosystems through the
        Missouri Botanical Garden: http://mbgnet.mobot.org
Information on biomes: www.blueplanetbiomes.org (a student-generated website)
Information on the jigsaw technique: www.jigsaw.org
Institute for Learning Innovation: http://ilinet.org

44

Using the Museum as a Resource for Learning Science

Jennifer D. Adams

INTRODUCTION

Approach the American Museum of Natural History on any given school day, and you are bound to see the grand buildings that make up this institution circled by a line of yellow school buses. Inside the museum, the halls are filled with excited students in a flurry of activity; some are running from diorama to diorama marveling at the exotic stuffed animals, others are diligently making careful observations and taking notes, some are gathered around the Teaching Volunteers' touch carts manipulating and learning about some of the objects behind glass, while others are more concerned about the location of the nearest bathroom and what time is lunch. Each day, close to 5,000 students and their teachers visit the museum.

Since the founding of the museum in 1869, the museum has been a central part of informal education in New York City. It has always been a primary goal of the museum to bring to the public view the work of scientists that takes place behind the scenes. If you were to ask anybody who was raised and schooled in New York City, undoubtedly they will have visited the museum as a student. The dinosaur fossils, the large blue whale, weighing yourself on the moon, and visiting the stars in the planetarium are all memories of the museum that people carry through a lifetime.

While many halls have been changed and updated, the sense of wonder and curiosity about our natural world that happens during a visit to the museum has remained constant. In this chapter, I will examine the unique characteristics of learning in a museum as they play out in the American

Museum of Natural History, followed by tips from museum educators for making the most of a visit to a museum.

WHAT MAKES LEARNING IN A MUSEUM UNIQUE?

John Falk (2001) describes learning in a museum as a *free-choice*, which is described as a self-paced learning environment that allows the learner to pursue his or her own interests and inquiries. Learning in a museum allows learners to participate in their own education in ways quite different than in a traditional classroom. In the museum, the learner immerses him or herself in scientific questions and investigations that allow the learner to be involved on the periphery of the scientific work that occurs behind the walls. In accordance with Jean Lave and Etienne Wenger's (1991) theory of legitimate peripheral participation, learners in a museum can become "apprentices" in the work of the scientists, as they are able to conduct expeditions in the halls and make observations in a similar way that scientists do in the field. This supports the National Research Council's idea of involving the public in the work of science—science as a human endeavor (Melba and Abraham 2002).

Lisa Roberts (1997) notes that there has been a shift from using the term *education* in describing a museum setting to using terms such as *learning*, *experience*, and *meaning making*. These terms convey the effective nature of the museum experience. She also refers to museum educator Raymond Stites from a 1963 article where he refuted the idea of museums as a leisure activity with frivolous end results, citing the "intellectual motivation" and "cerebral spiritual significance" of the museum visit. Mihaly Csikszentmihalyi and Kim Hermanson (1995) describe a "flow" state that visitors may experience in museums, a state of complete engagement where a visitor loses track of time.

Learning in museums allows for individual differences in interest and learning style and takes into account the social interactions involved in the nature of learning. This social inclination can also describe the meaning making that occurs with the interaction between learners and the objects in a museum (Leinhardt and Crowley 2002).

CHARACTERISTICS OF OBJECTS IN A MUSEUM

Gaea Leinhardt and Ken Crowley (2002) describe objects in a museum as example-based learning and enumerate four attributes of objects in museums that make them "unique nodes for ideas and their elaboration: resolution and density of information, scale, authenticity, and value."

1. *Resolution and density of information* refer to the realness of the object as opposed to the two-dimensional representation of a photo. This allows the viewer to see and experience detail, for example, the iridescence of feathers in a bird, the coldness of a stone, or the smell of an elephant in a zoo. A

striking object in the Hall of Planet Earth is the banded iron rock—a glistening black rock with multiple horizons of rust-red streaking across. This rock contains thousands of years' worth of information about our Earth's early atmosphere. Students can feel the coolness of the stone, they can count the red lines, and they can use a ruler to measure the intervals of red and black as they inquire about the Earth's changing atmosphere before the existence of complex living organisms. This aesthetically pleasing rock is both resolute and dense with information for both scientists and students. It is atmospheric information permanently etched in stone. More than a picture in a textbook, this rock is a tactile connection to the distant past.

2. Objects in a museum, if not the actual objects, are replicas carefully designed to depict an extant object—life in actual *scale*. Both children and adults alike can marvel at the size of a dinosaur fossil as they walk under it and up a ramp to get an eye-to-eye view. In the American Museum of Natural History's Fossil Halls, these life-sized fossils and casts are displayed in such a way that walking through the halls is like walking through the evolutionary history of vertebrates. Scale of size and scale of time—the experience of walking through the halls of the fourth floor makes concrete millions of years of history of vertebrate life on Earth.

3. *Authenticity* "exists in the interaction between specific objects and our history and culture" (Leinhardt and Crowley 2002). This allows us to view the objects in the context of our personal, everyday lives (and make sense accordingly). While the term *authenticity* is used to describe everyday objects that are often displayed in history and art museums, I believe that objects of science in a museum convey a degree of authenticity in that they can (1) connect us to our natural history—the history shared with all life on Earth; and (2) connect us to the objects and tools of authentic science. One of my favorite displays in the Rose Center's Hall of the Universe is a case with various small living and nonliving objects (a seashell, a piece of iron, and a beetle's carapace, among other things) with the following statement: "We Are Stardust. Every atom of oxygen in our lungs, of carbon in our muscles, of calcium in our bones, of iron in our blood—was created inside a star before Earth was born."

4. Finally, *value* refers not only to the monetary value of certain objects, but also to the cultural importance that objects hold. The Willamette meteorite is an example of an object that holds both scientific and cultural importance. As a meteorite, it provides important clues about the origins and chemical content of our universe. It is also a sacred object to the Clackamas tribe of western Oregon, who named the meteorite "Tomanowos," a revered spiritual being, and used the rainwater that collected in its pockmarks for ritual purposes. Every year, the Rose Center for Earth and Space where the meteorite is displayed is closed for a day as members of the Clackamas travel from far to visit Tomanowos and continue their tradition of interaction with this sacred object.

These four factors speak to how objects in a museum can support learning and allow the learner to experience the object—whether thinking about how the object was collected, imagining the behavior of an object when it was alive, or asking questions about the object—and personalize his or her experience with the object.

THE OBJECT AND THE VIEWER

While the objects in a museum maintain these characteristics, it is not the objects themselves that provide the unique learning experience, but the *interactions*—those of the object and learner, between learners, and of the learner and the space—that make the learning experience. Learners coming to a museum are bringing hosts of personal experience that will influence how and what they learn from the objects on display. Using the diorama as an example, each viewer may focus on one part of the diorama—one may hone in on the plants because she is a gardener, while another may scrutinize the background scenery because he likes art. George Hein and Mary Alexander (1998) and other museum researchers often use constructivist and discovery learning theories when describing the nature of learning in a museum. Both of these pedagogies place the learner at the center and emphasize the importance of social interaction, experience, and meaning making in the learning process. The role of the museum educator, teacher, or other adult is to convert the many dynamic, vivid, and motivating experiences into opportunities that promote learning. What follows are some tips to promote learning in a museum.

TIPS FOR USING THE MUSEUM AS A RESOURCE FOR LEARNING SCIENCE

Integrate Trips to the Museum in the Classroom Curriculum

Include visits to the museum as a part of the curriculum. Plan pre- and postvisit classroom activities in order to make a seamless transition from the museum learning experience to the classroom. For example, during a unit on classification, students could use a trip to the museum to find out more about a particular group of organisms, including the diversity of the group and the different environments that individual species could inhabit. The previsit lessons could include activities and discussions about characteristics of organisms and the nature of taxonomy, the museum trip could be framed like a scientific expedition to look for a particular group of organisms, and the postvisit activity could be a discussion of "findings" and lead into further inquiry about, for example, animal characteristics or adaptations (many other topics can stem from posttrip discussions). Perhaps students could be encouraged to visit the museum on their own for further research.

Practice Skills before the First Field Trip

Allow students to practice inquiry, observing, and recording skills before their trip to the museum. The museum is a very attractive place with many distractions, so it would be beneficial to have students in the practice of recording and observing before they visit the museum for the first time. One helpful exercise is having students look at a picture of a scene from a nature magazine. You could begin to ask them questions like "What do you see?" or "How do you know what season it is?" or "What do you think is going to happen next?" to get them to think and discuss beyond the scene that is presented. You could also bring objects like acorns, leaves, shells, and so on—stuff that you or your students collect—and begin to ask them to draw, label, categorize, and ask questions about what they are observing. You can develop an object inquiry sheet that encourages students to jot down objective (What do you see?) and subjective (What does it remind you of? What do you think?) observations, and provides a space for them to draw their object. Take them on a mini-expedition around the school. If inside the building, they could use their observations to learn as much about the fourth floor as possible, or they could go outside and observe the creatures that live around their school building. These exercises are a few that could get students in the habit of observing, recording, and asking questions—skills that will make a trip to the museum a more valuable learning experience.

Utilize Online Resources

In an effort to make museums more accessible to a wider audience, many museums have good websites that provide not only information about visiting the museum, but also information about the science that takes place behind the scenes in the museum. Ranging from floor plans and background information on exhibits to teacher's guides, curriculum resources, and articles about using museums and informal spaces for learning, these online resources are helpful in planning a visit to the museum and they can provide a wealth of background information about the various topics covered. This could prove a valuable resource in general curriculum planning.

Visit Ahead of Time

If possible, visit the museum ahead of time. This could be done during presemester planning or during holidays. Scope out the halls and displays that you wish to focus on. Decide how and where you will group your students to facilitate their learning experience—remember that during a field trip, museum halls will be crowded with other classes, and the acoustics may not be conducive for large-group instruction. You may want to have your students form working groups ahead of time (they may be already in groups if the trip is a part of an ongoing unit) and brief them on their museum activity before the trip. Once in the museum, you may want to locate a relatively quiet corner where you could group the students and give last-minute advisement.

Less Is More

The amount of information in a museum can be overwhelming. Rather than trying to see as much as possible in a day, pick one hall that specifically links to your curriculum. Encourage your students to really observe and record what they are seeing. You may want to provide them with a few tools to help their observations such as a flashlight, a ruler, and colored pencils.

Frame Distinct Guiding Questions and Goals for the Trip

A guiding question is a great way to focus a field trip. The museum field trip could be framed as an expedition with students going on a search for clues or answers to inquiries that they may have generated before the trip. Guiding questions also model how scientists do their research starting with observations leading to questions and a search for answers that often lead to more questions.

Allow Time for Free Exploration

Because a trip to the museum may be novel to students (even if they've been before, there is always something new to see or old to see again), it is important to allow students to freely explore some places of personal interest to encourage the sense of curiosity and wonder that a museum often inspires. This can promote a spirit of lifelong learning and the sense of the museum as a place for personal enjoyment.

REFERENCES

Csikszentmihalyi, Mihaly, and Kim Hermanson. 1995. Intrinsic Motivation in Museums: What Makes Visitors Want to Learn? *Museum News* 74 (3): 34–37, 59–61.

Falk, John. 2001. *Free Choice Science Education: How we Learn Science Outside of School.* New York: Teacher's College Press.

Hein, George, and Mary Alexander. 1998. *Museums: Places of Learning.* Washington, DC: American Association of Museums.

Lave, Jean, and Etienne Wenger. 1991. *Situated Learning: Legitimate Peripheral Participation.* Cambridge: Cambridge University Press.

Leinhardt, Gaea, and Ken Crowley. 2002. Objects of Learning, Objects of Talk: Changing Minds in Museums. In *Perspectives on Object-Centered Learning in Museums,* edited by Scott G. Paris, 302–24. Mahwah, NJ: Lawrence Erlbaum.

Melba, Leah M., and Linda M. Abraham. 2002. Science Education in U.S. Natural History Museums: a Historical Perspective. *Science & Education* 11:45–54.

Roberts, Lisa. 1997. *From Knowledge to Narrative: Educators and the Changing Museum.* Washington, DC: Smithsonian Institution Press.

ADDITIONAL RESOURCES

Suggested Reading

Roschelle, Jeremy. 1995. *Learning in Interactive Environments: Prior Knowledge and New Experience.* www.astc.org/resource/educator/priorknw.htm.

Websites

Articles written by and for educators about using museums as a resource for teaching and learning can be found at Musings: www.amnh.org/learn/musings/
Education activities related to permanent and temporary exhibitions can be found at the American Museum of Natural History Resources for Learning page: www.amnh.org/education/resources/index.php

45

Using Field Trip Experiences to Further the Participation in and Learning of Science

Kimberly Lebak

The use of informal learning centers, museums, interactive science centers, nature centers, and field study centers has greatly increased in number in recent years (e.g., Falk and Dierking 2000). As the popularity of informal learning centers has grown, their use for field trips by science teachers has also grown (Gilbert and Priest 1997). The National Research Council recognizes that "the classroom is a limited environment. The school science program must extend beyond the walls of the school to the resources of the community" (1996, 45). The *National Science Education Standards* note that informal learning centers "can contribute greatly to the understanding of science and encourage students to further their interests outside of school" (National Research Council 1996, 45). However, despite the increase in the use of informal learning centers as supplements to formal science education and the opportunities for promoting science learning (Griffin and Symington 1997), informal learning centers have a history of independence and autonomy (Bybee 2001) from the formal education system. Typically, a field trip consists of a one-day excursion to a location away from the classroom with little connection between classroom learning and topics studied at an informal learning center. The potential for learning through such partnerships has not yet been maximized.

Through this chapter, I describe how a partnership between an informal learning center teacher and a classroom teacher provides opportunities for maximizing student science learning during a unit of study on ecosystems. I describe the partnership between Gina, a fourth–sixth-grade science teacher, and me, a director of the Outdoor Classroom, a nature center. Through Gina and my collaboration, students are able to engage in real-world science learning through a field trip that supports the learning in Gina's classroom.

FORMING A PARTNERSHIP

Gina teaches science at a small independent school. During May 2004, she was planning a science unit on ecosystems and wanted students to go on a field trip as part of the unit of study. When she called to book the field trip for her fifth-grade science students, she asked that the field trip be an introduction to her unit of study. To truly make this a meaningful instructional experience for Gina and her students, I met with Gina and two science teachers from the Outdoor Classroom to best design the field trip with Gina's learning goals for her students in mind. Gina had distinct learning objectives for the field trip. She articulated how she wanted students to learn that all organisms in an ecosystem are interdependent. Gina also wanted students to recognize that we must take care of our ecosystems. Through our planning session, we discussed Gina's goals for the trip, her students' knowledge regarding ecosystems, and the opportunities for learning at the Outdoor Classroom. We also discussed the differences between the traditional learning structure of her classroom and the outdoor environment of the Outdoor Classroom. With Gina's goals articulated and the learning opportunities of the Outdoor Classroom outlined, we formulated a plan for the day that included students making predictions about what plants and animals live in given ecosystems and then taking the role of real scientists looking for evidence of their predictions that lead to a greater understanding of the interdependency of species within ecosystems. The initial meeting was instrumental in getting all instructional stakeholders aligned upon the goals of the day. Furthermore, as a result of our planning, Gina knew how to prepare her students in the science classroom for the field trip.

LEARNING OUTSIDE OF THE CLASSROOM

When Gina and her students arrive at the Outdoor Classroom, I introduce them to ecosystems through a prediction exercise for students to determine what plants and animals may live in a given ecosystem. In groups, students are handed an envelope with twenty cards of plants and animals and an ecosystem poster. The poster is divided into four sections representing four ecosystems, field, pond, stream, and woods. Students must decide in which ecosystem each plant or animal would be found. Although the exercise initially seems easy for fifth-grade science students, the students soon recognize that the exercise is raising more questions than answers. In the following excerpt Jordon, Kerry, and Austin struggle to place their cards, containing the names of plants and animals, on the ecosystem poster.

Speaker	Discourse
Jordon	This is a picture of a weird kind of grass.
Kerry	Grass is the field.
Jordon	Grass can go in the woods.
Austin	Where should we put it?
Kerry	I don't know.
Austin	Teacher, teacher, I don't know where to put the grass card?

The complexity of the interdependence of species is a very difficult concept for students to understand. However, this short exercise focuses the students on the exploration to come. Before we start the exploration, I ask students to tell me about the exercise.

Sam articulates his frustration: "We found that we may find plants and animals in more than one place. Some animals might sleep in one place and travel somewhere to find food. How do you know where something lives?"

"Let's see if we can find out," I reply.

EXPLORING THE POND

At the pond, the students become scientists, looking for evidence to confirm or refute their predictions. I give students tubs, bug jars, nets, test tubes, and magnifying glasses to explore the pond. Students are instructed to observe, draw, and identify all signs of life. I explain to students that all their findings will be used for two future activities, developing a food web and deciding on the health of the water source. The novelty of the pond setting could easily become a distraction for students used to learning in the confines of a classroom setting. However, through the planning session, Gina and I had articulated our goals to each other for students' participation and learning. We had also discussed the differences between the classroom setting and the outdoor setting. Therefore, during the field trip, our roles as teachers are already defined. Together, we are able to work with small groups of students to keep them focused upon the learning activity and objectives of the field trip. The following vignette illustrates how Gina and I work with Shara and Nina to facilitate their identification of the dragonfly nymph.

Speaker	Discourse	Gestures
Shara	Teacher, teacher.	
Gina	What do you have, what do you have? Look at him swimming. What kind of bug is it?	Peers at the water.
Nina	It's mean looking.	

Gina	Well, take some time and make some observations and record them. Use the pictures over there to help you decide.	
Shara	OK.	Shara draws while Nina begins to look through the reference materials.
Gina	Did you figure out what it is?	
Nina	A dragonfly nymph	
Kim	Yes. What do think the dragonfly nymph will become?	
Shara	I don't know.	
Gina	Look around. Do you see anything that the nymph may become?	Points in the direction of a dragonfly.
Nina	A dragonfly, really?	
Gina	Yes.	
Shara	Wow, the nymph looks too mean to become something so pretty.	

Identifying species in the ecosystem is the first step to discovering what plants and animals live in the pond ecosystem. However, Gina's objective was for students to see the interdependence within an ecosystem. Therefore, her interactions with her students support that goal. In the following excerpt, Gina uses Sam's and Ramon's findings to help them not only identify the water boatman in the water sample but also identify the algae that helps support the life of the water boatman.

Speaker	**Dialogue**	**Gestures**
Ramon	What are those?	Points to the insect in the tub.
Gina	Can you identify these? Here are our reference materials. Try to identify it.	
Sam	Are these plants?	Pointing to the tub.
Gina	Can you also use the plant material in your sample to help with the identification? What does the plant material tell you about this pond?	
Sam	That it is a food source.	Ramon and Sam begin to look in the reference book.
Gina	Right.	
Sam	Look, this is water boatman and it eats algae right here in the tub.	

Gina   Great work. Think of all
                the food that you are
                looking at. It is like
                appetizers for other animals.

As students continue to explore the pond ecosystem, they notice that life at the pond is not just isolated to the water. They find tracks of deer, evidence of beavers, turtles, frogs, red-winged blackbirds, dragonflies, and two water snakes. Students also observe different types of plants growing on the water's edge. When Gina spots a turkey buzzard flying around the pond, Gina and I are able to introduce the role of decomposers in an ecosystem.

Speaker	Dialogue	Gestures
Gina	Look at the hawk.	Points to the bird in the sky.
Kim	That's a turkey buzzard. That's one thing on the food web I did not talk about. The turkey buzzard is a scavenger.	
Sam	That means it eats dead things.	
Gina	Very good.	
Ramon	He looks for things that have died.	
Gina	So there is something dead over there.	
Tristin	Yes.	
Kim	Yes, there is something that has passed on over there.	
Gina	What could be dead over there?	
Ramon	Beaver.	
Sam	Deer.	
Kim	Yes, at this time of year, there could be a fawn that has died.	
Tristin	Oh, yuck.	
Kim	This is the time when the mothers are having the fawns.	
Gina	This is something to think about on your food webs.	

MAKING CONNECTIONS

After the exploration, students work in groups to develop a food web based upon their findings. As students talk through the making of a food web, they discuss not only what plants and animals live in a pond ecosystem but also how each member of the ecosystem is connected. Based upon their findings, the students identify producers, consumers, and decomposers within the ecosystem. Furthermore, through the development of the food web, students see the feeding relationships among organisms. They discover that the food web can be very complex as many species share the same area.

As Sam concludes, "Not only do the animals and plants that live in the pond need the water, but we found evidence of deer that travel to the pond to drink."

At the end of the creation of the food web, Gina and I ask students if they believe the pond is healthy, that is, free of any pollutants. Kirsten responds that she thinks the pond is healthy but is unsure how to support that belief. I introduce the concept of bioindicators. Students learn that water quality researchers often sample macroinvertebrate populations to monitor the water quality of a water source. Through a game, students learn which environmental stressors impact different macroinvertebrate populations. Students then compare their findings from the pond exploration with what they have just learned through the game. The students conclude, based upon their findings, that indeed the pond water is healthy. One student comments that it is great that the water is healthy because so many plants and animals within and near the pond ecosystem depend upon the water in the pond for survival.

EXTENDING LEARNING

Back in the classroom, the field trip experience provides the background for Gina to talk about different ecosystems and what things effect the continuation or the destruction of ecosystems. Students then use the information they had learned through the field experience at the Outdoor Classroom to design their own ecosystem project. For their projects, students design an experiment to answer a question they had formulated related to some aspect of something they had seen or heard during the field trip. The students developed a variety of different projects. The ecosystem projects demonstrate how students transfer the knowledge and skills learned through the field trip to their own scientific exploration. Tristan and Ramon questioned what pollutants could impact the lives of microinvertebrates living in a body of water. Through their project, they demonstrated how runoff silt, fertilizer, garbage, and car oil negatively impact the lives of the macroinvertebrates that live in water sources. Kristen determined the health of different water sources in her community. She accumulated samples of many water sources in her area and tested them for pollutants. She collected information from bioindicators to develop a chart of her results.

SUCCESSFUL FIELD TRIP EXPERIENCES

Field trip experiences at an informal learning center have the potential to promote science learning in an authentic manner. For Gina's students, the lesson on ecosystems moved outside the confines of the classroom environment to a real-world scientific exploration of an ecosystem.

When the students first arrived at the Outdoor Classroom, they struggled through the initial exercise to locate where plants and animals could live. They did not recognize the dependence of plants and animals on one another and a healthy environment. Through the one-day experience, students were able to identify different plants and animals in a pond ecosystem, find evidence of animals that may not live in the pond but rely on a pond for their own survival, and discover how to use bioindicators to determine the health of the pond ecosystem. The field experience provided opportunities for Gina's students to be real scientists, exploring their world, recording data, and using data to draw conclusions. The scientific skills learned through the day were transferable to future learning as demonstrated by the projects presented in the classroom.

This experience was meaningful for the students because Gina and I consciously became partners in the participation in and learning of science during the ecosystem unit. Most important to the success of this field trip was a clear understanding by all instructional stakeholders of the objectives for science learning. Through my understanding of Gina's goals for the trip, we were able to design the experience, using the resources of the nature center, to truly support Gina's teaching of a science unit on ecosystems. Gina and I both recognized that teaching is a complex task during an outdoor science field trip when students are experiencing new phenomena in a classroom without walls. Through our partnership, we were able to facilitate those goals through our interactions with students. As a result, students gained scientific knowledge and skills about pond ecosystems that directly transferred to future science participation and learning.

REFERENCES

Bybee, Rodger. 2001. Achieving Scientific Literacy: Strategies for Insuring that Free-Choice Science Education Complements National Formal Science Education Effort. In *Free-Choice Science Education: How We Learn Science Outside of School*, edited by John Falk, 44–63. New York: Teachers College Press.

Falk, John, and Lynn Dierking. 2000. *Learning from Museums: Visitor Experiences and the Making of Meaning*. Lanham, MD: AltaMira.

Gilbert, John, and Mary Priest. 1997. Models and Discourse: A Primary School Science Class Visit to a Museum. *Science Education* 81:749–62.

Griffin, Janette, and David Symington. 1997. Moving from Task-Oriented to Learning-Oriented Strategies on School Excursions to Museums. *Science Education* 81:763–79.

National Research Council. 1996. *National Science Education Standards*. Washington, DC: National Academy Press.

46

Building Adult Advocacy in Science: A Girl Scout and Science Museum Collaboration

Dale McCreedy

> I discovered that "hey I can do this" and I wanted other people to discover that too.
>
> ——Virginia, Girl Scout leader and National
> Science Partnership (NSP) trainer

With these words, a science-shy Girl Scout leader reflects on her experience in an informal science program for girls. Prior to her engagement in this program as an adult female volunteer, "Virginia" was an outsider—an outsider to science, an outsider to teaching, and even, to some degree, an outsider to Girl Scouting. Through her participation in a new community formed by the National Science Partnership for Girl Scout Councils and Science Museums (NSP), Virginia developed a new identity as an advocate for science, girls, and Girl Scouting.

Virginia has now been an advocate for girls and science for fifteen years, guiding science activities for 2–3 Girl Scout troops a year, each consisting of 8–15 girls, with additional science program offerings to younger Brownie Girl Scouts. Virginia has also facilitated many workshops for other leaders who have, in turn, led their own troops in science activities. What hidden potential this woman's story suggests—that informal experiences can open the door to science learning for adults who then may open the door to informal science learning experiences for girls!

INFORMAL AND COMMUNITY-BASED LEARNING

There is increasing national interest in learning that takes place before and after the school day and outside the classroom. Often called *informal* or *free-choice learning*, out-of-school time is characterized as self-directed and as free from prerequisites and assessment, and frequently includes social or group interactions as part of the experience. Opportunities for informal science education are readily available, especially at science museums, zoos, and aquaria; but may also be found in youth groups, clubs, and community-based organizations with family-oriented activities. A multitude of hobbies and other individual interests may also be classified as informal science learning.

This chapter attempts to identify and share the benefits and opportunities that arise from informal learning environments and community-based collaborations that support science learning. Inherent in these programs is the recognition that participating adults, often untrained or inexperienced at science or education, have the potential to be powerful advocates for children's science appreciation and achievement. If we wish to enhance children's science experiences, as well as science education nationwide, we need to better understand how to support the adults who have nontraditional opportunities to mediate children's learning.

The program described below is illustrative of the potential that exists in many available science programs offered through informal and/or community-based settings. I share my experience as the museum-based project leader at the Franklin Institute Science Museum, and thus I attempt to represent neither the full scope of science opportunities that are offered through Girl Scouts (nationally or locally) nor the full scope of those offered through other museums and science centers.

UNIQUE ELEMENTS OF GIRL SCOUTS AND NSP

The National Science Partnership for Girl Scouts and Science Museums (NSP) is a national collaboration between the Franklin Institute Science Museum and Girl Scouts of the USA, funded by the National Science Foundation to address science avoidance behavior in adults critical to girls—Girl Scout Leaders. Since 1988, NSP has grown from a small, local pilot effort to a national initiative established in more than seventy sites across the United States. NSP provides science activity kits and training for adult volunteers that directly correspond to badge requirements for Brownie and Junior Girl Scouts, and uses a train-the-trainer model for dissemination within a council site. The NSP involved more than 11,500 leaders and 130,000 girls during the period of federal funding from 1988 to 1996 and is still active today—more than a decade later—in sites across the United States.

I have found it useful to think about programs like the NSP as potential *communities of practice* as described in 1991 by Jean Lave and Etienne Wenger. These authors describe a community of practice as having three critical elements that have the potential to engage participants: (1) a network of people who are committed to (2) a common domain, goal, or purpose, and share (3) particular skills and practices to support the community's mission. The NSP, as well as the Girl Scouts more generally, represent a unique community of practice, with strategies for engaging and supporting adults, even science-shy ones, in science teaching and learning.

GIRL SCOUT COMMUNITY OF PRACTICE

Girl Scouts is the world's largest organization dedicated to helping all girls everywhere build character and gain skills for success in the real world. The purpose, or *domain*, of Girl Scouting is "to inspire girls with the highest ideals of character, conduct, patriotism, and service that they may become happy and resourceful citizens" (Girl Scouts of the USA 1998–2005c). The *community of people* in Girl Scouts includes girls, ages 5–17, and adults. There are more than 300 local Girl Scout councils nationwide, which together constitute nearly 4 million members, of whom 2.8 million are girl members and 986,000 are adult members, primarily volunteers (Girl Scouts of the USA 1998–2005a). These influential adults, perhaps neither teachers nor parents, have the potential to play significant roles in helping to shape the choices a girl makes about her achievement and future.

The *shared skills and practice* of Girl Scouts center around the Girl Scout Promise and Law and the following four program goals: to develop girls to their full potential; to help them to relate to others with increasing understanding, skill, and respect; to develop values to guide their actions and provide the foundation for sound decision making; and to contribute to the improvement of society through their abilities, leadership skills, and cooperation with others (Girl Scouts of the USA 1998–2005b). Girls earn badges or patches developed around the goals listed above after fulfilling the necessary requirements for each. Science is only one aspect of the Girl Scout program.

THE NATIONAL SCIENCE PARTNERSHIP
COMMUNITY OF PRACTICE

NSP began at a time when there was growing national concern about girls falling behind in math and science, and a long history of underrepresentation of women pursuing science coursework and science careers. National as well as local Girl Scout data suggested that only 2–5 percent of all badges earned by girls were in science. When asked to name the single thing that would help them to conduct better troop activities, the leaders' top three responses were training, ways to improve their knowledge base, and more ideas for activities. NSP met this need by offering adults resources and training opportunities that support the completion of Brownie and Junior Girl Scout (first- through sixth-grade) badge requirements in the sciences. However, the success of the program depends on whether the adult volunteer women serving as Girl

Scout Leaders, and potentially influential adults in children's lives, cultivate their own science interest and so are able to engage girls in hands-on science activities.

The NSP community of practice *domain* encompasses a commitment to the mission of Girl Scouting, but specifies a further commitment to support girls' and adults' experience and achievement in science. The *community of people* in NSP is layered, with groups that operate at the national, regional, and local levels. Nationally, Girl Scouts of the USA works with the Franklin Institute to offer NSP across the country. In Philadelphia, for example, there is a core group of adults from a number of Girl Scout councils in the region who collaborate in planning and shaping program offerings and sharing resources and practices.

NSP *shared practices and resources* are embedded in existing Girl Scout program structures. There are seven hands-on science kits filled with activities, explanations, and materials specifically linked to the Girl Scout badge requirements; leader workshops designed to increase the science knowledge and self-confidence of troop leaders; and partnerships between science-strong institutions and Girl Scout councils. Leaders obtain an NSP kit by participating in a two-hour workshop in which they engage in hands-on science activities representative of the most exciting and most difficult activities included in a specific kit. They subsequently return to their troop and engage in science exploration with their girls over the course of 5–7 troop meetings. Along with this repertoire of activities, NSP has symbols (logos, badges), vocabulary, and themes, including "Science is fun," "It is OK not to know," "Girls can do science," and "Leaders are critical influences." All become a common ground upon which participants share practices and become members in the NSP community of practice.

NSP IMPACT

In 1996, an evaluation of the NSP adult participants, undertaken by the Education Resources Group, showed that during the project's development, there were increases in leader confidence in facilitating girls' science activities and cultivating girls' enthusiasm, positive attitudes, and long-lasting interest in science. In 2003 and 2005, my research focusing on leaders who had participated for more than ten years not only confirmed these results but also documented ways in which participation had been reinforcing and, for some science-shy leaders, transformative (McCreedy 2003, 2005). Although not conclusive, interviews of high school–aged NSP participants suggested long-term impact on girls' continued interest in science and involvement in Girl Scout–related science activities. In 2000, NSP was recognized by the Department of Education as an effective program in encouraging gender equity in science.

The elements of the community of practice also provide a useful framework for understanding the impact of NSP. The *mission or domain* of NSP—the commitment to the support of adults' and girls' science learning—has been compelling for all participants, whether museums, councils, leaders, or girls:

> I see the benefits of our participation in this program every time I lead a session with one of the science kits. I see leaders who were at first apprehensive about science or the material, finishing excited and more self-confident about offering science programs to their troops. (Science Museum staff member)

> NSP provided a springboard for our council to offer many science-related activities for girls and adults. (Girl Scout council staff member)

> When I told the girls we were going to work on the Science Sleuth Badge, their reaction was "yuk." After the first week, they couldn't wait for more. It changed their whole idea about science and this is something we definitely should strive for. (Junior Girl Scout leader)

The NSP science activities, workshops, and materials were critical to the development of *shared practices and skills.*

> I would have skipped right over [the badge] without the training. I would not have done the research to provide the girls with this type of program. . . . I feel it is very important that they're introduced to science the fun way that you introduced science to us. (Girl Scout leader)

> We all learned and had fun. The constellations are something we still talk about and look for. The girls still tell me what the moon's shape is and [our] life in space discussion amazed me. (Brownie Girl Scout leader)

Participants found different aspects of the *community of people* engaging. For some, being among other leaders who shared membership in the Girl Scout community was sustaining in and of itself, while others became engaged by the participation of those with particular interests in science or gender issues and for whom Girl Scout membership was secondary. In either case, participants shared in the camaraderie of being with others who felt excited about NSP.

> It is nice to be with people who share interests, get excited [about science].

> NSP gives us a chance to be educators of girls in a relaxed atmosphere. It takes the barriers away.

CULTIVATING AND SUSTAINING MEMBERS OF THE NSP COMMUNITY

How do informal science programs cultivate and sustain adult members, like Virginia, whose involvement and commitment grow and influence the development of the community? Virginia, one of the earliest NSP leaders, has evolved over fifteen years into an expanded role as a facilitator of kit work-shops for other leaders, council coordinator of NSP, and promoter of her council's outdoor education efforts. With her new sense of herself as a science learner and teacher, she is now empowered to think of herself as a critical part of the community's future. She has also developed membership in over-lapping communities, which play a part in sustaining her membership in the NSP community of practice.

Multimembership

We are all members of many communities. However, some communities intersect in ways that seem to leverage the impact of another. Virginia's identity as a science practitioner was the result of multimembership in Girl Scouting and NSP. Girl Scouts provided the critical community of practice that nurtured Virginia's engagement with science learning and advocacy for girls in science. Involvement as a supportive mother with her daughter's troop first exposed her to the Girl Scout community that, through NSP, led to the acquisition of science skills and resources, increasing commitment to science education as an NSP facilitator, and increasing advocacy for Girl Scouting.

> I would do trainings with leaders and I'd have found leaders in there that were in my situation—they had never done science in school...and they were homemakers or secretaries or something very traditionally feminine....And there were also women who were GS leaders who were engineers and biochemists for Exxon and both of them got a lot out of the workshops because when you are...you've been fighting your way through school learning to be a biochemist to compete with the men you still don't know how to teach girls science, hands-on fun at a very simple level and a couple of them said afterwards "you know I would have done these things in a much more complicated way that would have lost the girls' interest...this is cool." And again, I heard "this is cool" and I said yes, it was worthwhile for me to do this, to go through what I go through to get to a training, at a time when I have all three kids at home and a husband that travels.

As an "old-timer" in NSP, Virginia was able to offer legitimacy to others as they became engaged. She invited in newcomers by sharing her own "nonscience" background and by encouraging future members of NSP to become advocates for girls and science and collaborators within the Girl Scout organization.

Evolution of the Community

Just as the individual learns and evolves as a result of her participation in the community, so, too, the community evolves as a result of her participation and influence. For example, Virginia used and continues to use the NSP science program with her troops of Brownie and Junior Girl Scouts, as initially designed. However, in adapting the kits for use as a troop fundraiser for her Cadette and Senior Girl Scout troops, she adjusted the program structure so that older girls now take an active part in providing science activities to younger girls. This is a variation to the original structure that focused only on leaders, but it expands the community of empowered learners to include young women moving toward adulthood. And so, through Virginia's involvement and commitment to Girl Scouts and NSP, girls who were Brownie Girl Scouts a decade ago are now members in the NSP community of practice, as shown by their participation in NSP training workshops and their leadership of younger girls in science.

Expanding the Boundaries

NSP is structurally defined by its very specific and targeted overlap with the Girl Scout community of practice, including its all-girl environment and its badge-related curriculum. However, this program is not exclusive in its expectation that participating adults—often outside of their field—will share enthusiasm and cultivate interest in science. By attempting to understand active participation in NSP, I began to understand the ways in which informal environments, often thought to be marginal to dominant educational beliefs and practices, could offer outsiders an entrée into science learning and teaching. For those lacking backgrounds in or historically fearful of science, informal programs like NSP can be a second chance, a new way of understanding science without the barriers of the past. Participation provides opportunities to develop a relationship with science, and a new identity as someone who enjoys science, engages in science activities, and opens doors to science learning. Once transformed, these individuals can become (1) sustaining contributors to the evolution of the community that gave them a new identity, and (2) powerful advocates for the transformation of others.

There are many adults in children's lives, and most are neither trained scientists nor educators. The youth group leader, after-school care provider, child care worker, camp counselor, and parent—all are untapped resources, "crucial intermediaries" between children and science. If we want all children to reach their potential, all aspects of a child's educational experiences need to be considered and integrated. This requires that we have a better understanding of the ways in which adults from many settings become engaged and, ultimately, develop identities that allow them to open the door to science learning, in and out of school. Through NSP, it is clear that adults who sustain involvement in an informal community of practice centered on science learning can open the door to children's involvement in the larger science community. Think of the potential!

REFERENCES

Education Resources Group. 1996. *Evaluation and Follow-Up of the National Science Partnership: 1993–4 and 1996*. Princeton, NJ: Author.

Girl Scouts of the USA. 1998–2005a. Facts. www.girlscouts.org/who_we_are/facts/.

———. 1998–2005b. Girl Scout Program. www.girlscouts.org/program/.

———. 1998–2005c. Who We Are. www.girlscouts.org/who_we_are/.

Lave, Jean, and Etienne Wenger. 1991. *Situated Learning: Legitimate Peripheral Participation*. New York: Cambridge University Press.

McCreedy, Dale. 2003. Educating Adult Females for Leadership Roles in an Informal Science Program for Girls. Ph.D. diss., University of Pennsylvania.

———. 2005. Engaging Adults as Advocates. *Curator* 48 (2): 159–76.

WEBSITES

American Association of Museums: www.aam-us.org

American Zoo and Aquarium Association: www.aza.org

Association of Science-Technology Centers: www.astc.org

Boys and Girls Clubs of America: www.bgca.org

Girl Scouts of the USA (National Site): www.girlscouts.org

Information about 4H Locations: www.4h-usa.org

National 4H site: www.fourhcouncil.edu

National Science Partnership (NSP): www.fi.edu/tfi/programs/nsp.html

National Wildlife Federation: www.nwf.org

YMCAs: www.ymca.net/index.jsp

YWCAs: www.ywca.org

47

Pitfalls in the Teaching of Evolution: Darwin, Finches, History

Catherine Milne

WHY HISTORY?

Joseph Schwab, who helped develop the Biological Sciences Curriculum Study (BSCS), argued that one needed to know the history of theories, not just the theories or facts, in order to understand that scientific theories evolve (Schwab 1978). Knowing a little about the history of a theory helps to make us more aware of the role of inquiry in the construction of science knowledge rather than seeing science as a collection of facts that need to be learned. Such an understanding makes a person less likely to accept simplistic explanations of theories and more likely to appreciate that competing arguments about the mechanisms of theories can exist even when the broad theory is accepted. One such theory is evolution.

WHY IS IT HELPFUL TO KNOW HISTORY?

I was sitting in a biology class, observing students who were working on a simulation of natural selection called "The Beaks of Finches." This required laboratory activity for students studying Living Environment (a biology course) involved using tools such as long-nosed pliers, chopsticks, and clothespins as analogies for a type of finch beak. Finches are small perching birds that belong to the Passeriformes (Latin for sparrow) order of birds. The finches, upon which this simulation activity is based, come from the Galápagos Islands, a small archipelago in the Pacific Ocean sitting on the equator about 600 miles west of South America. Charles Darwin made the Galápagos Islands, and the animals living there, famous in his books, especially *The Voyage of the*

Beagle and *The Origin of Species* ([1845] 2003), where he argued for evolution by natural selection as the mechanism for formation of new species of living things. Darwin struggled, as we do today, to define exactly what a species could be, and he did not use the idea developed later of separate species as "noninterbreeding groups" separated by heredity or environment (Coyne 1994). Instead Darwin argued, "No one definition has yet satisfied all naturalists; yet every naturalist knows vaguely what he means when he speaks of a species" (Darwin [1845] 2003, 572).

HOW DOES THE SIMULATION WORK?

The simulation required three items (see Figure 47.1), each tool representing a different finch beak. To begin, each student predicts her tool's potential for picking up small seeds from a large dish. The tools had been randomly assigned because "evolution is random and therefore the assignment of beaks is random." However, the structure of beaks in offspring is not random but is based on heredity as each finch inherits genes from its parents for beak structure. The simulation includes three rounds (see Figure 47.2).

Every round (see Figure 47.2) effectively asks the question "How does a certain beak shape affect the competitiveness of a finch?" Only the number of seeds collected measured "survival." In the body of the activity, teachers are advised to assert that all fourteen species of Darwin's finches differ from each other in body size and/or beak size and shape, implying that finch species are easy to identify. However, recent studies indicate that these species are not as differentiated as the information provided with this simulation claims and that specific finch species can be very difficult to identify unless the island from which a finch is collected or observed is known (Grant 1986).

Observing this simulation being enacted posed various questions: how central were Darwin's finches to the development of his theory of evolution and its mechanism, natural selection? What did this simulation imply about natural selection, and were these implications supported by historical records? Was there a historical source for the tool analogy on which the simulation was based? How genuine was this exercise for understanding inquiry?

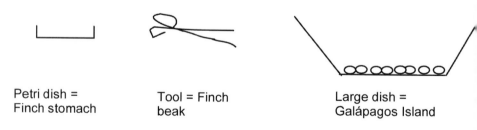

Petri dish =
Finch stomach

Tool = Finch
beak

Large dish =
Galápagos Island

FIGURE 47.1   The items needed for the activity "beaks of finches."

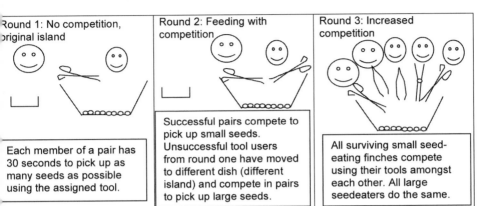

Round 1: No competition, original island	Round 2: Feeding with competition	Round 3: Increased competition
Each member of a pair has 30 seconds to pick up as many seeds as possible using the assigned tool.	Successful pairs compete to pick up small seeds. Unsuccessful tool users from round one have moved to different dish (different island) and compete in pairs to pick up large seeds.	All surviving small seed-eating finches compete using their tools amongst each other. All large seedeaters do the same.

FIGURE 47.2   The structure of the simulation.

STARTING AT THE BEGINNING

The first place to look for a record of the role of Darwin's finches in Darwin's thinking is in his published works, especially *The Voyage of the Beagle*[1] and *The Origin of Species*.[2] The edition of *Voyage* that I have was published in 1845, nine years after Darwin had returned to England after his five-year global voyage.

Darwin was originally appointed as a gentleman companion to the *Beagle's* captain, Robert Fitzroy, who desired a companion of similar social background for conversation so he would not go mad and commit suicide as the previous captain had done (Larson 2001). Darwin had just finished his studies at Cambridge University but had no immediate plans for employment (Sís 2003). On the long voyage as the *Beagle* completed its task of surveying South America for the British Navy, Darwin made extensive observations of the places visited, including the Galápagos Islands, which today are part of Ecuador. When Darwin and the *Beagle* visited the Galápagos Islands in 1835, he noted the recent arrival of a permanent human settlement because prior to this no humans had lived there. Today several towns support the tourism and fishing industries on the Galápagos Islands, and the Ecuadorian government has been challenged to resolve simmering confrontations between the fishing industry and national park rangers and conservationists (Nordling 2004).

DARWIN AND THE VARIETY OF LIFE ON THE GALÁPAGOS

The archipelago is a little world within itself, or rather a satellite attached to America, whence it has derived a few stray colonists, and has received the general character of its indigenous productions.

—Darwin ([1845] 2003, 389)

Darwin believed that all the organisms on the Galápagos originally came from the Americas. Plate tectonics provides a mechanism for the appearance of the Galápagos Islands, which are volcanic, suggesting that the original organisms on the islands must have come from the nearest land mass, America, just as Darwin proposed. A consequence of the volcanic origins of the islands is that from about 5 million years ago, when the first Galápagos Island emerged, the number of islands available to finches has increased over time with a similar increase in the number of species of Darwin's finches (Grant and Grant 2002). Although the finches were not the only group that Darwin found unusual and impressive, they clearly intrigued him:

> The remaining land-birds form a most singular group of finches, related to each other in the structure of their beaks, short tails, form of body, and plumage: there are thirteen species.... The most curious fact is the perfect gradation in the size of the beaks in the different species of Geospiza.... Seeing this gradation and diversity of structure in one small, intimately related group of birds, one might really fancy that from an original paucity of birds in this archipelago, one species had been taken and modified for different ends. (Darwin [1845] 2003, 391)

Geospiza is the genus name for the ground and cactus finches from the Galápagos Islands that were identified by the ornithologist James Gould (Sulloway 1982). *Genus* (plural, genera) is a level of classification for living things that is more general than species and more specific than the next major level, family, a hierarchical structure for organizing living things originally proposed by Carl Linnaeus (1707–1778). Put simply, the name of a species occurs in Latin, made up of two parts, hence *binomial nomenclature*: the first name identifies the genus, and the second part is a descriptive name for the species. For example, the sharp-beaked ground finch is called *Geospiza difficilis*. *Geospiza* is the name of the genus, and *difficilis* is Latin for difficult or troublesome (perhaps referring to the challenges associated with identifying this bird). Since the eighteenth century, this approach to nomenclature has been the accepted way to communicate the identity of species of plants and animals across languages and nationalities.

In conclusion to this section of the *Beagle*'s voyage, Darwin commented on the apparent diversity of species:

> [S]everal of the islands possess their own species of tortise [sic], mocking thrush, finches and numerous plants. These species having the same general habits, occupying analogous situations and obviously fitting the same place on the natural economy of this archipelago that strikes me with wonder. (Darwin [1845] 2003, 409)

Use of the term "natural economy" is interesting because it suggests that in 1835, Darwin was still reconciling his thinking about evolution with the conventional ideas of the time. Since the mid-seventeenth century, naturalists had observed that many lower animals and plants produced enormous numbers of offspring, but almost all of them died before reaching an age at which they could reproduce. This futile production of offspring had to be reconciled with "God's love for His Creation"—how could he let so many die?—and explained against the fact that most animal and plant populations were fairly stable. Linnaeus in his *Oeconomia Naturae* (Hestmark 2000) suggested that each creature had its allotted place in nature, having been assigned its peculiar food and geographic range or niche. Thus competition with other creatures was avoided, ensuring harmony and plenty. At this time in the development of natural science, natural economy and binomial nomenclature reinforced this static concept of niche.

Darwin's theory of evolution was to overturn the theory of natural economy. He and William Wallace, who studied organisms on the Malay Peninsula, jointly released their proposals for a theory of evolution to the Linnaean Society in 1858.[3] In *The Origin of Species*, Darwin largely ignores the finches, using many other examples including the domestication of animals by humans to provide evidence for his theory. It was left to other researchers in the next century, including David Lack, Robert Bowman, and Peter and Rosemary Grant, visiting the Galápagos Islands sometimes for long periods of time and examining hundreds (possibly thousands) of different finch skins collected by numerous visitors to these islands, to develop theories about the speciation of Darwin's finches. It was in the publications of some of these researchers that the tool analogy for the beaks of finches first appeared.

DO TEXTBOOKS REPRESENT DARWIN'S FINCHES DIFFERENTLY?

Peter Grant, writing about Darwin's finches, argued that textbooks often oversimplify the complexities of the "evolutionary processes they illustrate, the ambiguities of the evidence, and the differences of opinion among biologists about just how these birds evolved" (1981, 653). Although Grant argues that Darwin's observations of the finches "contributed substantially" to the development of his theory of evolution, Darwin's own writings, especially *The Origin of Species*, do not support this assertion. Perhaps it is not that observing finches provided Darwin with insight into evolution. Rather, Darwin's emerging ideas of evolution provided him with insight into the complexities of the finches' emergence.

However, there is no doubt that Darwin's finches have become associated with contemporary representations of evolution, especially in high school biology textbooks. For example, George Johnson and Peter Raven in *Holt Biology* (2004) state that Darwin "collected 31 specimens of finches from 3 islands" (290), but fail to mention that Darwin did not label the collected finch skins with the islands from which the finches were collected (Sulloway

1982). Only in hindsight did he recognize the possible importance of knowing which finches came from which islands, providing insight into how ideas and experiences are interwoven in scientific inquiry. Darwin's insight was further reinforced when Johnson and Raven claim that Darwin's hypothesis was framed this way: "Galapagos finches evolved from an ancestral type and changes in the finches occurred as different populations accumulated adaptations to different food sources" (277–78). According to them, in 1938 David Lack, who introduced the collective term "Darwin's finches," tested Darwin's hypothesis but found little evidence to support it observing finches with different beaks eating the same types of seeds. When Peter and Rosemary Grant conducted their studies in drought years, they observed that particular beak shapes provided some finch genera with selective advantage. However, when Lack revisited his data in the 1940s (long before Grant; see Lack 1947), he reconsidered his initial arguments and argued for the importance of geographical isolation and competition to the formation of finch species on the Galápagos (see Figure 47.3).

According to the simulation it was competition that caused some finches to die out and some to thrive, which is consistent with Lack's argument for finch speciation. However, is that the only possible explanation for the appearance and behavior of the finches? The simulation implies each type of finch is restricted to a specific type of seed, but studies initially by Lack (1947) and then by Robert Bowman (1963) indicated that each species of finch tended to eat a range of foods. For example, although the diet of ground finches consisted mainly of seeds, insects constituted about 20 percent of their diet. Bowman argued that it was not competition but adaptations for food getting that caused the evolution of Darwin's finches. Today's researchers, however, consider that both competition and environment are important (Grant and Grant 2002).

BUT WHERE DID THOSE TOOLS COME FROM?

In the activity, students used tools to simulate different types of beaks. Bowman (1963) extensively studied the beaks of finches and proposed the tool analogy that was later represented in Grant (1986) to provide a sense of beak structure and underlying musculature (see Table 47.1).

The power of this tool analogy is evident in its adoption in scientific and general interest publications, and as the basis for "The Beaks of Finches" simulation. However, the tool analogy as developed by Bowman (1963) was designed to give readers a sense of the different beaks found in the different *genera* of finches, not *species* as the activity the students were completing implied. For example, both ground and cactus finches have beaks like heavy-duty linesman's pliers. An awareness of the history of this analogy suggests that if educators continue to use tools as an analogy for the beaks of finches, it might be beneficial for students to discuss how the analogy is and is not consistent with their beaks and how the types of food (e.g., types of seeds used

Evolution of Darwin's finches by *adaptive radiation* – a model
　　　Ancestor finch species arrives in the Galápagos Archipelago from the Americas.

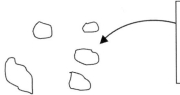

Ancestral finch arrives in Galápagos Archipelago and settles an island. This happens more than once as studies suggest that the most recent ancestor was an insect eater rather than a seedeater as was originally thought (Schluter, 2000).

2.　　Geographic isolation and natural selection. Population of finches grows and carrying capacity of island is reached. Finches disperse to other islands, are isolated on those islands, and tend to diverge as selection pressure is enacted. As finches respond to different environments they develop more suitable traits.

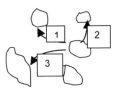

3.　　Responding to environmental factors, populations of finches disperse to islands already populated by a different group of finches. What will happen? This is the question that was addressed implicitly by the simulation, "The Beaks of Finches."

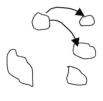

If enough differences have evolved in the finches while they have been isolated from each other, population mix factors such as different niche selection and the development of differential behaviors, such as song and plumage might be significant enough to isolate species from each other even when they are living on the same island. This indicates that new species have formed. However, Grant and Grant (2002) claim evidence from their study of finches that for some populations of finches geographical isolation is enough to cause the formation of new species.

FIGURE 47.3　A model for speciation.

and only seeds) used in the simulation are also consistent and not consistent with the food of finches on the Galápagos.

IMPLICATIONS

　　Tools for "The Beaks of Finches" might help students to make connections between the characteristics of organisms and resource use and the implications of that for natural selection and speciation. However, there is a need to encourage students to ask questions such as what are the limitations of this analogy, and how does the analogy help us to make sense of the processes that drive evolution? There is a need to be wary of accepting analogies such as "tools for beaks" and not asking these questions, because without such

TABLE 47.1

Finch Genus	Types of Finches	Number of Finch Species	Tool Bowman Proposed for each Genus of Finch
Geospiza	Ground and Cactus finches	6	Heavy-duty linesman's pliers
Camarhynchus	Tree finches	3	High-leverage diagonal pliers
Cactospiza	Woodpecker and mangrove finches	2	Long chain-nose pliers
Platyspiza	Vegetarian finch	1	Parrot-head gripping pliers
Pinaroloxias	Cocus Island finch	1	Curved needle-nose pliers
Certhidea	Warbler finch	1	Needle-nose pliers

questions learners might be led to believe that these analogies are "truthful" representations of both the organisms and processes being studied.

Why know history? History helps us to understand how theory develops and to appreciate the role inquiry plays in theory development. Darwin did not observe his finches on the Galápagos Islands and immediately formulate his theory of evolution with its mechanism of natural selection. Instead, his observations on Galápagos and elsewhere spurred his thinking and further study that extended over twenty years. You might ask where is the inquiry in the simulation "The Beaks of Finches," and how can this activity be made more inquiry based? You might also ask questions such as "Using similar materials, how could you model the ancestor species arriving on a specific island? How could you model geographic isolation? If competition is not the only factor driving natural selection, how could you model other factors?" Hopefully, this analysis has provided an inkling of how to unfold the debates that currently inform the development of evolutionary theory and provided a richer sense of the strengths and limitations of analogies for learning science.

NOTES

1. Reprint of the second edition originally published in 1845 called *Journal of Researches into the Natural History and Geology of the Countries Visited during the Voyage of the H.M.S. Beagle round the World*, originally published by John Murray.

2. Reprint of the first edition of *The Origin of Species*.

3. You might wonder why Wallace tends to be less well remembered than Darwin. It could be because Darwin developed his arguments more fully than Wallace or because of Wallace's later unwillingness to place humans in the same evolutionary schema as all other living things.

REFERENCES

Bowman, Robert I. 1963. Evolutionary Patterns in Darwin's Finches. *Occasional Papers of the California Academy of Sciences* 44:107–40.

Coyne, Jerry A. 1994. Ernst Mayr and the Origin of Species. *Evolution* 48:19–30.

Darwin, Charles. [1845] 2003. *The Origin of Species and the Voyage of the Beagle.* New York: Everyman's Library, Alfred A. Knopf. (This edition reprints the text of the second edition of *The Voyage of the Beagle* and the first edition of *The Origin of Species.*)

Grant, Peter R. 1981. Speciation and the Adaptive Radiation of Darwin's Finches. *American Scientist* 69:653–63.

———. 1986. *Ecology and Evolution of Darwin's Finches.* Princeton, NJ: Princeton University Press.

Grant, Peter R., and B. Rosemary Grant. 2002. Adaptive Radiation of Darwin's Finches. *American Scientist* 90:130–40.

Hestmark, Geir. 2000. Oeconomia Naturae. L. *Nature* 405 (6782): 19–20.

Johnson, George B., and Peter H. Raven. 2004. *Holt Biology.* New York: Holt Rinehart and Winston.

Lack, David. 1947. *Darwin's Finches.* Cambridge: Cambridge University Press.

Larson, Edward J. 2001. *Evolution's Workshop: God and Science on the Galapagos Islands.* New York: Basic Books.

Nordling, Linda. 2004. Crisis in Darwin's Paradise. *New Scientist* 184 (2475): 6–7.

Schwab, Joseph. 1978. *Science, Curriculum and Liberal Education.* Chicago: University of Chicago Press.

Sís, Peter. 2003. *The Tree of Life: A Book Depicting the Life of Charles Darwin, Naturalist, Geologist and Thinker.* New York: Frances Foster Books.

Sulloway, Frank J. 1982. Darwin and His Finches: The Evolution of a Legend. *Journal of the History of Biology* 15:1–53.

48

Teaching Electrochemistry

Pamela J. Garnett

WHAT IS ELECTROCHEMISTRY?

Some chemical reactions produce electricity, and conversely sometimes electricity can produce chemical reactions. This interconversion of chemical and electrical energy is called *electrochemistry*. In electrochemical reactions, electrons are transferred from one substance to another.

When they are controlled, electrochemical reactions are of great benefit to society. Chemical reactions that produce electricity result in the flow of electrons in wires and of ions in ionic solutions. These reactions underpin the construction of electrochemical (galvanic) cells such as the flashlight battery, car battery, and fuel cell. The resulting flow of electrons moves through a metallic conductor to a device that uses electrical energy, for example a globe, starter motor, or fuel ignition system.

In contrast, sometimes electricity or electrical energy from an external source can cause chemical reactions. The process of converting electrical energy into chemical energy is called *electrolysis*. Examples include the conversion of gold ions into gold plate on jewelry ($Au^+ + e^- \rightarrow Au$) or silver ions into silver plate on a nickel spoon ($Ag^+ + e^- \rightarrow Ag$). Electrical energy is also used in the refining of copper and the extraction of aluminium metal from alumina. Electrons are forced through a metallic wire by an external source of electrical energy, such as a battery. The ends of the wire connect to electrodes that are immersed in an electrolyte. At the interface where the electrode meets the electrolyte, oxidation-reduction reactions are forced to occur.

THE DEVELOPMENT OF OUR UNDERSTANDING
OF ELECTROCHEMISTRY

Our understanding of electrochemistry has evolved over time. At first scientists guessed that positive particles were the current in metallic conductors, but now we know that very small, negatively charged electrons move throughout a network of "fixed," positively charged ions. In the past, oxidation reduction was defined in terms of the transfer of oxygen or hydrogen. Nowadays we more commonly use *electron transfer* and *changes in oxidation state* to define oxidation reduction, because these definitions are more inclusive of different contexts.

In addition, research into students' understanding of electrochemistry can be used to improve classroom practice. Accurate preconceptions of electric current and oxidation-reduction reactions are necessary to develop accepted understandings of electrochemical cells (galvanic cells) and electrolytic cells.

TEACHING TO IMPROVE STUDENTS' UNDERSTANDINGS
OF ELECTROCHEMISTRY

My teaching is informed by alternative conceptions research and by many years of experience. I view my time with students as an opportunity to maximize learning; and to achieve this outcome I structure, scaffold, and facilitate learning, encouraging students to enjoy, learn, think about, and be engaged in the practice of chemistry. My students are in their final two years of high school, competing for limited places at university. Our school has an extremely strong science and mathematics tradition, with over half of our girls entering science-based university degree programs.

The following section includes a discussion of my approach to teaching electric current and oxidation reduction, and lesson notes to introduce the concepts of electrochemical and electrolytic cells. The "notes" are teacher directed, and students are expected to listen, think, answer and ask questions, and take notes from the whiteboard during these introductory lessons. Following this, they complete follow-up activities that require group practical work, participate in student-student discussions, and complete written exercises for feedback and consolidation.

ELECTRIC CURRENT

Often electricity is referred to as "the flow of electrons" with no mention of current in solutions. Working with this primitive concept of current creates difficulties for students of electrochemistry. The outcome is they think that electrons move in ionic solutions and create mechanisms for this movement.

For students of both physics and chemistry, using conventional current, or the "flow of positiveness," creates difficulties. Students tend to compartmentalize their knowledge, saying that electricity in chemistry and physics is different because electrons flow in the opposite direction. Using this outdated convention results in considerable confusion for many students, and the use of conventional current needs to be questioned.

Opportunities to develop the concept of electric current arise when teaching the structure and properties of metals, ionic solids, and covalent substances. The model for metallic structures, a lattice of positive ions and negatively charged mobile electrons, provides the opportunity to stress that current in metals is the movement of delocalized electrons.

Similarly, ionic solids that have been dissolved in water to form a solution of mobile positive and negative ions should feature in discussions of current in aqueous solutions. To explain this conductivity, attention needs to be paid to the source of electrical energy (the battery); the flow of electrons from the negative terminal of the applied voltage through a metallic conductor; the fact that reactions that gain and lose electrons occur at respective electrodes; and the movement of positive and negative ions in opposite directions in solutions.

Some covalent molecular substances such as acids also break up into ions, and these provide additional opportunities to discuss conductivity in solutions.

OXIDATION AND REDUCTION

Several models are used to define oxidation-reduction reactions. Originally, *oxidation* referred to reactions with oxygen that resulted in the formation of oxides and to reactions in which the reacting species gained oxygen. Clearly, these definitions are inadequate in some instances. Consider the reaction of magnesium with oxygen to form magnesium oxide, $2Mg + O_2 \rightarrow 2MgO$. Students correctly think that magnesium is oxidized because it gains oxygen, but they do not think that oxygen is reduced. By focusing solely on the addition of oxygen, they erroneously believe that oxidation occurs independently from reduction.

Some students erroneously think the reaction of a soluble carbonate with an acid, $CO_3^{2-} + 2H^+ \rightarrow H_2O + CO_2$, is an oxidation-reduction reaction because hydrogen has gained oxygen. Others make the mistake of thinking it is an oxidation-reduction reaction because there has been an electron transfer between the carbonate and hydrogen ions. They think that the carbonate has "lost" two electrons to the positively charged hydrogen ions.

To identify oxidation-reduction equations with certainty, changes in the oxidation state of the atoms need to be used. Using several different models causes confusion for learners, especially as some earlier models only apply to specific situations and therefore have limited application. Given that pre-conceptions are resistant to change, previously learned inadequate models have the potential to be generalized to situations in which they do not apply and to interfere with subsequent, more sophisticated learning. The reason for using several models for the same concept needs to be clearly explained to the learner. For example, is it because a "simple" model is used to make conceptually difficult information accessible to young students, or is it to show how ideas in science evolve over time?

When using changes in oxidation states to identify oxidation-reduction equations, a common error is to use the total charge on a polyatomic ion instead of the oxidation state of each atom. For example, in the equation $H^+ + OH^- \rightarrow H_2O$, some students say that the oxidation number of the hydroxide ion is -1 and that it goes to zero when water is formed. By not calculating the oxidation states of the hydrogen and oxygen atoms, students incorrectly identify this as an oxidation-reduction equation.

Another common error students make is to assign elements the oxidation state of their ion. For example, students frequently assign magnesium metal an oxidation state of $+2$ instead of zero for the element. If teachers are aware of these learning errors, they are able to take more care with the language they use to discuss these concepts and to select examples such as these to challenge erroneous ideas. These strategies lead to improved teaching and learning.

ELECTROCHEMICAL CELLS

A four-step lesson plan to introduce the concept of electrochemical cells is provided below. The steps address essential prerequisite knowledge and draw from alternative conceptions research.

1. Understanding current: revise the definition of electric current by referring to conductivity tests students have previously performed. This ensures students include the movement of ions in solutions as well as electron flow in metals.

2. Addressing misconceptions about why current flows: use the demonstration outlined in Figures 48.1 and 48.2 to illustrate that electrochemical reactions occur because (1) they are energetically favorable, and (2) there is a competition for electrons.

   After a short period of time, a reaction is observed in Beaker 1: the zinc turns black and the blue solution fades in color. There is no reaction in

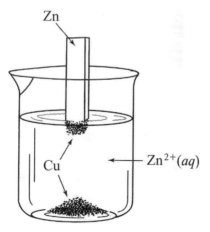

FIGURE 48.1

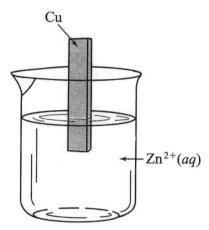

FIGURE 48.2

Beaker 2. Ask students, "Is the Cu, Zn charged? Is the solution charged? Why does Cu^{2+} react with Zn?" These questions directly challenge erroneous notions that the electrodes are charged, the solution has an overall charge, there is an excess of electrons on zinc, and the reason current flows is because electrons are moving from a region of high density (on the zinc) to low density (on the copper).

3. Developing the idea of competition for electrons by introducing the Standard Reduction Potential Table: ask students to write the ionic equation for the previous reaction and discuss and label Cu^{2+} as the electron acceptor and Zn as the electron donor, Cu^{2+} (acceptor) $+ Zn$ (donor) $\rightarrow Cu + Zn^{2+}$. Then write the half equations in order of reduction potential.

$$Cu^{2+} + 2^{e-} \rightarrow Cu$$

$$Zn^{2+} + 2^{e-} \rightarrow Zn$$

After this discussion, show students a Standard Reduction Potential Table.

4. Introducing electrochemical cells by talking about the idea of converting chemical energy into electrical energy. Using the $Cu^{2+}/Cu//Zn^{2+}/Zn$ cell as a demonstration and a diagram on the whiteboard (Figure 48.3), pose the following questions and involve students in discussion.

   • Is there a way of setting up Cu in Zn^{2+} so we can produce a current? This question will be rhetorical if students have no previous experience of electrochemical cells. Introduce the idea of separating the reagents because they react spontaneously, and lead students to the idea of half cells. Show students the demonstration cell.

   • Which way will the electrons move in the wires? Based on the previous discussion, students should answer, "From zinc to copper." Add that copper ions will move to the copper metal to gain electrons from the metal surface, resulting in the formation of more copper.

   • Why do the electrons move? Students should answer in terms of energy difference or competition for electrons.

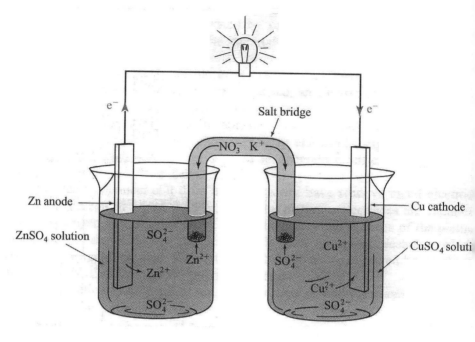

FIGURE 48.3

- Which way will the positive copper and zinc ions move? Why? My explanation is that copper ions are converted to Cu metal so additional positive ions move to take their place, maintaining electrical neutrality. This may be developed further by including the movement of zinc ions.

- Which way will the negative nitrate ions move? Why? My explanation is that when electrons flow from the zinc through the wire, negative ions move to maintain neutrality. In summary, negative ions move in the same direction as the electrons, either clockwise or anticlockwise depending on how the cell is set up. Positive ions move in the opposite direction.

   After these key ideas are developed, terminology such as *electrodes*, and the *anode* and *cathode* as the respective sites for oxidation and reduction, may be introduced. Also, if necessary, it may be mentioned that the anode can be labeled negative because it supplies electrons to the external circuit and the cathode can be labeled positive because it accepts electrons. Stress that the electrodes are not charged but by convention merely labeled negative and positive respectively. For preference, I omit this labeling because students get confused if they associate it with the movement of ions in the cell.

   Explaining the function of the salt bridge as "allowing for the flow of ions to maintain electrical neutrality" serves to address erroneous ideas about charge building up in half cells and electrons flowing in electrolytes. The need for a soluble salt such as ammonium nitrate to prevent precipitation reactions occurring in the half cells can also be addressed.

Several lessons follow in which students develop their ideas further, calculating the predicted E° for cells and relating this to the need to separate the reagents to prevent a spontaneous reaction (usually by using half cells). They also perform experiments and test their ideas in a range of different applications.

TEACHING ELECTROLYTIC CELLS

I introduce the electrolytic cell by talking about the idea of converting electrical energy into chemical energy and by referring to some uses of these cells. The major idea to develop is that an electric current can be used to produce a chemical reaction. I demonstrate the electrolysis of aqueous copper iodide (Figure 48.4) with graphite electrodes and at the same time diagrammatically represent the cell on the whiteboard. During a teacher-led question-and-answer session, I add detail to the diagram to facilitate students' development of the concept.

The sequence of questions I use is as follows:

1. "Which way are the electrons moving?" Students' attention is drawn to the applied emf[1]/voltage and the positive and negative labeling of the terminals. The electron flow is marked on the diagram as emanating from the negative terminal, as by definition this supplies electrons to the circuit.

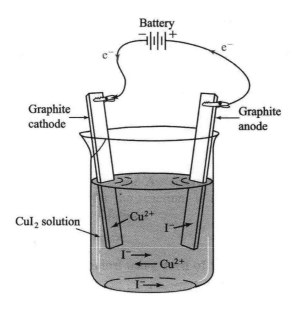

FIGURE 48.4

2. "Where will the oxidation half reaction take place, and where will reduction occur?" Following this questioning, students are able to identify the appropriate electrodes and hence the anode and cathode.

3. Next I focus on the movement of ions and ask students the direction of movement of positive ions and negative ions. As with the electrochemical cell, the negative ions move in either a clockwise or anticlockwise direction consistent with the electron flow, and the positive ions move in the opposite direction.

4. To determine the predicted half reactions at the anode and cathode, and the voltage that needs to be applied, students first list the species present at the anode and cathode. Then, using the Standard Reduction Potential Tables, they write down the possible anode and cathode reactions from the oxidation and reduction "sides" of the table. Next, they select the half reactions that result in the most positive $E°$, bearing in mind that the $E°$ will be negative or zero.

5. Next, students combine the anode and cathode half reactions to get the cell equation and the predicted $E°$. The implication of a negative voltage/$E°$ needs to be discussed in relation to the applied voltage.

NOTE

1. Electromotive force, also known as *potential difference*.

SUGGESTED READINGS

Garnett, Pamela J., Patrick J. Garnett, and David F. Treagust. 1990. Implications of Research on Students' Understanding of Electrochemistry for Improving Science Curricula and Classroom Practice. *International Journal of Science Education* 12:147–56.

Garnett, Pamela J., and David F. Treagust. 1992. Conceptual Difficulties Experienced by Senior High School Students of Electrochemistry: Electric Circuits and Oxidation Reduction Equations. *Journal of Research in Science Teaching* 29:121–42.

———. 1992. Conceptual Difficulties Experienced by Senior High School Students of Electrochemistry: Electrochemical (Galvanic) and Electrolytic Cells. *Journal of Research in Science Teaching* 29:1079–99.

Garnett, Patrick J., Pamela J. Garnett, and Mark W. Hackling. 1995. Students' Alternative Conceptions in Chemistry: A Review of Research and Implications for Teaching and Learning. *Studies in Science Education* 25:69–95.

Huddle, Penelope A., Margaret D. White, and Fiona Rogers. 2000. Using a Teaching Model to Correct Known Misconceptions in Electrochemistry. *Journal of Chemical Education* 77:104–10.

Ogude, Nthabiseng A., and John D. Bradley. 1994. Ionic Conduction and Electrical Neutrality in Operating Electrochemical Cells. *Journal of Chemical Education* 71:29–34.

Sanger, Michael J., and Thomas J. Greenbowe. 1999. An Analysis of College Chemistry Textbooks as Sources of Misconceptions and Errors in Electrochemistry. *Journal of Chemical Education* 76:853–60.

49

Demo Essays: Using an Old Trick to Teach in a New Way

Tracey Otieno

The science demonstration is a common feature in most high school science teachers' "bag of tricks." Teachers often use science demonstrations to inspire curiosity with a dramatic introduction to a topic, to present a discrepant event to raise awareness of a commonly held misconception, or to simply demonstrate a concept with a physical example. In spite of this versatility within the curriculum, demonstrations remain in the domain of "instruction" and are rarely used in assessing student understanding. Teachers communicate what is important about the subject area to their students through assessments (Farenga, Joyce, and Ness 2002). It is not only the teacher's selection of content material but also how they conduct their assessments that convey this message about what is relevant in science to students (Shepard 2000). Through incorporating demonstrations into assessment practices, my class and I were able to break out of traditional teacher-student roles and integrate scientific practices, not just content, into our assessments.

NECESSITY (AND SOMETIMES FAILURE) IS THE MOTHER OF INVENTION

Like many of the most influential scientific discoveries, many of my most effective science teaching innovations were accidental, or were attempts to recover after poor student performance on an exam. This was also true of the development of the science demonstration essay. Following the devastating results of a multiple-choice achievement test on gas laws, I gave my students a makeup exam. Because I was very troubled by the disparity between student performance on this achievement test and the knowledge and understanding they demonstrated during classroom interactions, I decided to try something I had never done before. Instead of giving another conventional exam, I did two demonstrations and asked the students to write a semistructured essay about them. The two demonstrations used were the can-crushing demo (where a soda can containing steam is inverted into an ice bath) and the demonstration where a hard-boiled egg gets drawn into a flask. An ongoing theme running through my class was the need to understand chemistry on three levels: the macroscopic, the molecular, and the symbolic (Gabel 1999). Consistent with this idea, the demo essays were structured such that in the first paragraph, students were to describe what they observed happening at the macroscopic level in the first demo. In their second paragraph, students were to explain the phenomena they observed by describing what was happening on the molecular level in the first demo. They repeated this pattern in their third and fourth paragraphs for the second demo. In the fifth paragraph, they were asked to compare the two demonstrations.

I was very apprehensive about giving this type of exam. I was afraid that students would be resistant to writing essays because in an average class, on traditional exam essay questions, about one third of the students would not even attempt an answer despite my promises of partial credit. Contrary to my expectations, however, some students were actually enthusiastic about writing these demonstration essays. One low-performing student who rarely participated in class exclaimed, "Finally!" when I described the assignment. Apparently this student loved to write, and I never knew or would have known that had I not attempted something different. After watching the demos, students began to write. I was astonished, as my normally noisy classroom grew silent with students concentrating on their writing.

INSIGHT INTO PROCESS AS WELL AS PRODUCT

The contrast between students' poor performance on the multiple-choice exam and the understanding of gas laws and kinetic molecular theory demonstrated in their essays was dramatic. It made me reflect on how difficult it is to design test questions that truly measure student understanding. If students misinterpret a question, they can get a wrong answer on a question while having a correct understanding of a concept. Alternatively, depending on how a question is crafted, students can give the appropriate response while harboring misconceptions that go undiagnosed by the test question. In an ideal world, teachers could assess student understanding by interviewing individual students, but this is logistically impossible. The demonstration essay is the next best thing. Demo essays allow teachers to gain clearer insight into students' thought processes and mental models. In demo essays, students have the opportunity to develop their own explanations for phenomena they are witnessing directly; therefore, there is little room for ambiguity or misinterpretation. Students have the chance to tell a story and develop explanations in their own words.

Some might question whether having students write a macroscopic description of what's happening in the demo has any scientific or educational value. If the overarching goals of science education are considered, there are things we want students of science to understand and be able to do beyond simply understanding scientific concepts. Certainly, any scientist would agree that making and recording observations are critical to the practice of science. Therefore, it is important that they appear on assessments in order to convey their significance to students.

In addition, I have found that having students begin with a description of a demonstration they just witnessed allows students who lack confidence in science (or in writing) to begin writing. Even if students are not sure about the molecular explanation for the demonstration, they can write about what they saw happen in the demo, and this is often enough to keep them writing. On the day of the first demo essay, one of the special education students in my class sat cross-armed in his seat, refusing to write. After reexplaining the assignment to this student, he exclaimed, "Oh!" and immediately began eagerly writing. Once he had written his description, he was confident enough to propose an explanation. His account of molecules "running for their lives," describing the steam escaping from the can (although anthropomorphized), indicates he had a useful starting "mental model" for the relationship between heat and kinetic energy.

ASSESSMENT: NOT AN ENDPOINT BUT A NEW BEGINNING

Reading my students' demo essays made me regret that this was the end of the gas laws unit. Their responses to the assignment allowed me to finally see how they conceptualized the behavior of gases, what they understood, and what misconceptions they held. Fortunately, on the day this first demo essay was conducted, I had a visitor in my classroom. He suggested that this should not be an endpoint, but a new departure. He proposed that instead of immediately moving on to another topic, I should make use of this assessment to further student learning by having students from different classes peer review each other's essays and create a final draft incorporating the comments of their peers.

The following day, I made copies of all of my students' essays, eliminating their names for anonymity. The essays of one class were distributed in the other, and vice versa. Before essays were distributed, we discussed as a class what a good essay should contain and collectively developed a rubric for rating each paragraph of the essay. Students read the work of their peers, scoring their essays using the rubric. Students took the responsibility of grading and critiquing each other's essays very seriously.

On the third day, students wrote the final draft of their demo essay. This process allowed the students to self-reflect, and allowed me to identify and address the misconceptions some students were harboring. If we had not gone through the revision process, only I would have been able to see their misconceptions, and there would have been no opportunity for the students to reflect on and correct their erroneous beliefs about why the can crushed and why the egg was drawn into the flask. This process of writing, reviewing, and revising transformed assessment from a judgmental, grade-producing device into a learning tool for the students and for me.

VARIATIONS ON A THEME

One of the beauties of the demo essay is its versatility. I have described here the first time I stumbled upon the idea of using demo essays; however, there were many other situations in which I made use of demo essays, and each time was different. I would not want anyone to assume that I am proposing the description of my experience as a formula for using demo essays. Therefore, I am listing some of many possible variations:

- Demo essays can be used with any science demonstration.
- The structure of the essay can be modified to meet the learning objectives of the class.
- Other scientific practices such as making predictions before the demo, posing new questions raised by the original demo, or proposing additional experiments one could use to determine answers to the new questions could be included.
- The follow-up activities can also be modified to suit the goals and time constraints faced by the class.
- Other possible follow-up activities could include student presentations, or having students conduct the demonstration and explain the science to another class.

REFERENCES

Farenga, Stephen J., Beverly A. Joyce, and Daniel Ness. 2002. Reaching the Zone of Optimal Learning: The Alignment of Curriculum, Instruction, and Assessment. In *Learning Science and the Science of Learning*, edited by Rodger W. Bybee, 51–62. Arlington, VA: NSTA Press.

Gabel, Dorothy. 1999. Improving Teaching and Learning through Chemistry Education Research: A Look to the Future. *Journal of Chemical Education* 76:548–53.

Shepard, Lorrie A. 2000. The Role of Assessment in a Learning Culture. *Educational Researcher* 29 (7): 4–14.

SUGGESTED READINGS

Atkin, Myron J. 2002. Using Assessment to Help Students Learn. In *Learning Science and the Science of Learning*, edited by Rodger W. Bybee, 97–103. Arlington, VA: NSTA Press.

Cox-Petersen, Anne M., and Joanne K. Olson. 2002. Assessing Student Learning. In *Learning Science and the Science of Learning*, edited by Rodger W. Bybee, 105–18. Arlington, VA: NSTA Press.

Kovac, Jeffrey, and Donna W. Sherwood. 2001. *Writing across the Chemistry Curriculum: An Instructor's Handbook*. Upper Saddle River, NJ: Prentice Hall.

50

The Sumatra Earthquake and Tsunami, December 26, 2004

*Dave Robison, Steve Kluge,
and Michael J. Smith*

The incredible media coverage of the earthquake and tsunami of December 26, 2004, provided the world with images provoking both incredible sadness at the tragic loss of life, and awe at the destructive energy released from a convulsion of Earth's crust. Earth scientists everywhere were immediately eager to study the earthquake that rattled the entire planet and generated the tsunami that destroyed so much on the rim of the Indian Ocean basin. As earth science teachers, we viewed the event as a teachable moment ripe with possibilities.

The relationship between the underlying geologic processes and the day-to-day lives of people was made instantly, obviously, and painfully clear. Here was an opportunity for our students to see firsthand the importance of earth scientists' efforts to understand and perhaps eventually predict the workings of the earth system. The ignorance of those happy but oblivious beachgoers who followed the receding sea before the arrival of the waves, and the story of the young girl who had learned the significance of such a recession (and saved many lives by warning others to leave the beach), drove home the value of personal understanding and preparedness. The irony of the instantaneous appearance of images and videos of the destruction, juxtaposed with the lack of warning to people who might have had hours to prepare, illustrated the fact that technology is limited by how effectively it is employed. Armed with actual data related to a real, familiar event, we were prepared to introduce our students to the concepts in seismology and plate tectonics through thoughtful and careful analysis.

BACKGROUND

The Sumatra–Andaman Islands earthquake of December 26, 2004, was the fourth largest earthquake in the world since 1900 and was the largest since the earthquake that struck Prince William Sound in Alaska on Good Friday in 1964. The Sumatra earthquake occurred as thrust-faulting on the interface of the India plate and the Burma plate. In a span of minutes, the faulting released elastic strains that had accumulated for centuries from on-going subduction of the India plate beneath the overriding Burma plate. Stein and Okal (2005) trace the underlying geologic causes of this tremendous earthquake back more than 120 million years ago, when the supercontinent Gondwanaland broke apart. Back then, a large landmass that we now recognize as the subcontinent of India split off from Antarctica and started its slow steady motion northward. About 50 million years ago, it collided with Asia, producing the Himalayas and the Tibetan plateau. Plate collision proceeds today as the Indian plate moves northward and forces pieces of Southeast Asia and China eastward.

Part of the boundary between the Indian plate extends along a trench (the Java Trench) on the west coast of Sumatra. In this region, an oceanic part of the Indian plate is being subducted beneath the Burma plate. The Burma plate is a microplate, or a small sliver between the India plate and the Sunda plate, which contains most of Southeast Asia. As the Indian plate is subducted, this creates a thrust zone along the plate boundary. Most of the time, there seems to be very little happening along this great (mega-) thrust fault that forms the plate boundary interface. However, there's really a lot going on. The Burma plate and Indian plate converge or move toward each other about 22 millimeters every year. Because the mega-thrust fault is locked, strain builds up on it. Sooner or later, the accumulated strain exceeds the frictional strength of the fault, and it slips as a massive earthquake like the one that occurred in December 2004 (Stein and Okal 2005).

The devastating tsunami was a direct consequence of the earthquake, which caused movement of the seafloor all along the length of rupture, displacing a huge volume of water and generating the tsunami wave. The vertical uplift could have been as much as several meters. In the open ocean, tsunami waves move very rapidly, 300–500 km/hour (about the speed of a jet airliner), and are often very small (a few centimeters). As the wave approaches the coast, it slows down and grows in height so that it can be many meters high when it strikes the coast.

CREATING THE ACTIVITY

Shortly after the December 26, 2004, earthquake, we downloaded a number of maps and original seismograms using the Global Earthquake Explorer1 (GEE), and found that working out the epicenter location and origin time from the seismograms yielded remarkably accurate results. From there on, we each worked in tandem, developing additional activities and supporting materials (web links, images, and our own web pages) that would introduce or reinforce understandings of the geologic principles involved and the connections between science, technology, and the lives of everyday people. In a steady stream of e-mails each built upon the other's ideas, this comprehensive classroom exercise achieved its present form. The activity consists of the following sections:

- Part 1: Students analyze actual seismograms, calculate P and S wave travel time differences, and determine the corresponding epicenter distances. This information is recorded in a table on the student exercise. Using the data, students then plot the epicenter distances on a map and triangulate the epicenter location, and finally work backwards to determine the origin time of the quake. Students that are careful every step of the way will be rewarded with a very accurate, pinpoint location of the epicenter on their maps, and with origin times within a few seconds of the actual time. Questions accompanying Part 1 lead students to examine the nature of the plate boundary east of Indonesia, including the study of maps and cross sections, web resources provided by the U.S. Geological Survey (USGS), as well as original material developed for this exercise.

- Part 2: Skills and concepts developed in Part 1 are reinforced as students analyze additional seismograms and plot the epicenter location on a larger scale map.

- Part 3: Students study the nature of tsunamis, and determine the location of the advancing waves over a period of several hours.

- Part 4: Students determine the average velocity of seismic waves that have traveled various distances, and are asked to reflect on the reasons for the observed discrepancy. This final activity provides an excellent anticipatory set for teachers wishing to further explore the nature of the Earth's interior and how this is revealed through the interpretation of seismic wave propagation.

- Part 5: This section contains a number of concluding questions regarding the human impact of the event. Students view video clips and still images, and are asked to review their data and explore ways in which various locations might have been spared some of the enormous loss of life.

Our downloadable exercise (available in both MS Word and PDF formats at www.regentsearthscience.com/tsunami.htm; Robison and Kluge 2005) includes seismograms, maps, seismic wave travel time charts, and a world tectonic map customized for the Indian Ocean basin.

REFLECTING ON THE ACTIVITY

To date, this lab has been used with hundreds of students. If anecdotal reports from teachers are accurate, the results have been outstanding. Teachers have noted the enthusiasm with which students have approached the lab, and are happy with the concepts that the students retain. The breadth of the lab provides several "kickoff" points for additional discussions and study of plate tectonics; the examination of the distance that volcanoes are from the surface expression of the plate boundary, the depth to which a plate must dive before magma is generated, and earthquakes produced by the rising magma.

Plotting tsunami travel times, watching associated videos, and discussing who had a chance to get out of the way (and even what "out of the way" means) brought the relationship between day-to-day human activity and the abstract idea of subducting plates into clear focus for our students. Taking it a step farther, we've been able to discuss the similarities and differences in the Pacific Northwest of the United States, illustrating it all with Earth Science Picture of the Day (EPODS, located at http://epod.usra.edu; Universities Space Research Association n.d.), recent articles regarding hazards of the Pacific Northwest, and personal photos and anecdotes.

For years, high school earth science students have used artificial or doctored seismograms to plot earthquakes that, as far as they're concerned, may or may not have ever happened. These "quakes" have no real connection to their lives, and as such the activity often devolves into the mere acquisition of a skill. What has impressed us with our activity is how students who work carefully can so accurately plot the position and calculate the origin time of a very real and current earthquake using real, undoctored data and a simple travel time graph that has been part of most high school earth science textbooks for many years.

REFERENCE

Stein, Seth, and Emile A. Okal. 2005. The 2004 Sumatra Earthquake and Indian Ocean Tsunami: What Happened and Why. *The Earth Scientist* 21 (2): 6–11.

SUGGESTED READINGS

Bolt, Bruce A. 2004. *Earthquakes.* 5th ed. New York: W. H. Freeman.

Brumbaugh, David S. 1999. *Earthquakes, Science and Society.* Upper Saddle River, NJ: Prentice Hall.

Levy, Matthys, and Mario Salvadori. 1995. *Why the Earth Quakes.* New York: W. W. Norton.

Prager, Ellen J. 2000. *Furious Earth.* New York: McGraw-Hill.

Robison, David, and Steve Kluge. 2005. The Great Sumatra Earthquake and Tsunami of December 2004: A Comprehensive Inquiry Based Classroom Exercise for High School Students. March. www.regentsearthscience.com/tsunami.htm.

Sieh, Kerry, and Simon LeVay. 1998. *The Earth in Turmoil.* New York: W. H. Freeman.

Stein, Seth, and Michael Wysession. 2004. *An Introduction to Seismology, Earthquakes and Earth Structure.* Malden, MA: Blackwell.

Universities Space Research Association. N.d. Earth Science Picture of the Day. http://epod.usra.edu.

U.S. Federal Emergency Management Association. 2005. *Tsunami Website.* www.fema
.gov/areyouready/tsunamis.shtm.

U.S. Geological Survey. 2005. *Earthquake Hazards Program.* http://earthquake.usgs
.gov.

Zelinga de Boer, Jelle, and Donald T. Sanders. 2005. *Earthquakes in Human History:
The Far-reaching Effects of Seismic Disruptions.* Princeton, NJ: Princeton University Press.

51

Investigating Volcanic Flows through Inquiry

Michael J. Smith and John B. Southard

Using simple fluids and materials, students explore factors that affect volcanic flows (viscosity, slope, magma temperature, and channelization). They explore the relationship between volume and surface area of a flow. The activity helps them to develop an understanding of the nature and hazards of lava flows, pyroclastic flows, and lahars. Students also design a controlled experiment, and learn how knowledge of earth science contributes to wise planning.

BACKGROUND INFORMATION

Fluids, in contrast to solids, are substances or materials that continue to change in shape for as long as they are acted upon by a deforming force. Fluids are of two kinds: liquids and gases. Liquids maintain a definite volume, whereas gases expand to fill their container. Mixtures of liquid or solid particles in a gas, and mixtures of solid particles in a liquid, also behave as fluids, unless the particles are so numerous that they form an interlocking network. Magmas (molten rock) are liquids. Magmas commonly contain solid particles in the form of crystals that grow suspended in the magma as it cools. Magmas, especially when they approach the Earth's surface, can contain gas bubbles as well, which rise slowly upward through the magma.

The two most significant physical properties of a fluid are its density and its viscosity. These two properties are not directly related. The density of a fluid is its mass per unit volume. The viscosity of a fluid describes its resistance to deformation. Here is a good "thought experiment" to give you a better understanding of the concept of viscosity. Lay a plate of glass horizontally on a table, smear it with a thick layer of a viscous liquid like honey, molasses, corn syrup, or motor oil, and then cover the layer of liquid with another plate of glass. Attach suction-cup handles to the top of the upper plate, and slide the upper plate horizontally, parallel to the lower plate. The layer of liquid between the plates undergoes a shearing motion. The greater the viscosity of the liquid, the more force you have to exert to keep the upper plate moving at a given speed. Magmas and lavas, although they are liquids, have very high viscosities, much higher even than everyday viscous liquids like those mentioned above. Students should not have the impression that magmas and lavas are as "flowy" as the liquids that are part of their everyday experience. Silica-rich magmas are of higher viscosity than silica-poor magmas.

As with any liquid, lavas flow downslope in response to the pull of gravity. A given volume of lava in a flow is acted upon by the downslope component of the force of gravity. Newton's second law of motion would tell you that the fluid should be accelerated in its downslope motion. The reason it instead flows at an almost constant speed is that a friction force develops at the bottom of the flow, which counterbalances the downslope driving force of gravity. Lavas flow faster on steeper slopes because the downslope force of gravity is greater. Lavas that flow in channels tend to move faster than lavas that flow as wide sheets, because the area of the base of the flow, where the retarding force of friction is generated, is less in relation to the volume of the flow, to which the force of gravity is proportional. For the same reason, thicker flows of lava tend to move faster than thinner flows.

Pyroclastic flows are a very different kind of flow that is sometimes caused by an explosive eruption. If the volcano emits a large volume of a concentrated mixture of gases, liquid droplets, and solid particles, the mixture can flow downslope like a dense liquid. The effective viscosity of such a mixture is much less than the viscosity of a lava, so the speed of movement is much greater. Speeds of pyroclastic flows can be over 100 meters per second, which is even faster than racing cars! The potential for destruction of plant and animal life, and of human habitation, is staggering. When pyroclastic flows finally stop, they sometimes become welded into solid rock as the still-hot material cools.

Lahars are a kind of debris flow whose solid materials are mainly or entirely volcanic ash. Debris flows are downslope-flowing mixtures of water and solid particles. If the concentration of solid particles is sufficiently high, greater than 40 to 50 percent by volume, the mixture of water and particles tends to flow like a viscous liquid. Because of the high concentration of solid particles, they cannot readily settle out to the bottom of the flow, as would be the case in an ordinary river flow, so the mixture can flow for long distances. If any of your students have had experience with concrete, they could appreciate that debris flows have a consistency something like that of fresh concrete that has just a bit too much water. Speeds of debris flows vary from very slow, no faster than a walk, to very fast, tens of meters per second. Areas susceptible to debris flows are those where slopes are steep and where loose sediment containing fine as well as coarse materials are common. Land surfaces with steep slopes and a mantle of weathered volcanic ash are particularly susceptible to debris flows.

Debris flows whose solid materials are mainly or entirely volcanic ash are called *lahars*. Lahars commonly develop during heavy rains some time (often a long time) after an explosive eruption has covered the land with a thick layer of volcanic ash. Lahars are especially common after weathering of the ash in a warm, humid climate has produced abundant fine clay material in the layer. Debris flows can start in various ways: runoff down slopes during and after especially heavy rainfall; collapse and sliding of steep, water-saturated slopes; or breakout of temporary lakes in terrain covered by volcanic ash. Debris flows, including lahars, tend to find their way into preexisting stream or river valleys, and when they finally come to a stop, the valley is filled with a deep layer of watery sediment. Entire villages can be buried almost instantly in this way. The village of Herculaneum, near the modern city of Naples, Italy, was deeply buried by a lahar that resulted from an eruption of Vesuvius two millennia ago.

In this activity, the students are asked to develop hypotheses and then design experiments to test those hypotheses. Commonly, science really works that way. Hypotheses come about by virtue of scientists' previous understanding, which might be based on an existing theory or on observational evidence already at hand. Sometimes, however, scientists just have a hunch or intuition that a certain new kind of experiment will reveal some information that opens up a new area of thought and research. Then the theoreticians have to scramble to develop new theories based on the new observational evidence, rather than the other way around!

Good design of an experiment is of critical importance in science. The experimenter needs to vary each of the variables or parameters that might have an effect on the process or phenomenon, or at least take them into account by making sure they are held constant. The values of each of those variables or parameters have to be chosen to cover the range of interest, and the number of values has to be chosen so that the "experimental matrix" (number of variables versus numbers of values of each variable) stays within workable limits. You might impress your students with the idea that, for example, with three or four variables and four or five values of each variable, they would end up having to perform anywhere from twelve (3×4) to twenty (4×5) individual experiments!

THE ACTIVITY

Goals

In this activity, students will do the following:

1. Measure and understand how volume, temperature, slope, and channelization affect the flow of fluid.

2. Apply an understanding of factors that control lava flows, pyroclastic flows, and lahars (mudflows).

3. Apply an understanding of topographic maps to predict lahar flow (mudflow) patterns from a given set of data.

4. Describe volcanic hazards associated with various kinds of flows.

5. Become aware of the benefits of earth science information in planning evacuations and making decisions.

6. Show understanding of the nature of scientific evidence and the concept of a controlled experiment.

Preparation, Materials Needed, and Teaching Tips

Spend most of one class period with the opener (to reveal student's initial conceptions) and the first stages of Part A (below). This will allow students to become familiar with the materials and brainstorm about possible experiments. You could review their designs that day and provide feedback the next time the class meets.

Part A

Encourage the students to think carefully about the design of the experiment to test their hypothesis. This includes the experimental setup or arrangement, the variables or phenomena to be observed, and the values of the variables to be used.

You will need a liter of a viscous (slow-flowing) fluid. The fluid should be one that changes its viscosity greatly when heated or cooled. Liquid soap works well and cleans up easily. Alternatives include molasses, corn syrup, shampoo, and glycerin.

Photocopy a piece of graph paper with centimeter grids onto overhead transparencies. Affix the transparencies to a manila folder. If you want a larger surface area with which to work, tape two transparency grids together and affix them to an opened manila folder. The folder provides support for some experiments. In lieu of overhead transparencies, laminate sheets of graph paper that have been affixed to manila folders are suitable. Eye droppers usually have a capacity of 1 milliliter or 1 cubic centimeter. Students will need books to create various slopes, and a source of hot water and ice to conduct experiments on the effects of temperature on viscosity.

Items needed for this experiment are as follows:

- Liquid soap (or a similarly viscous liquid; see above)
- Eye dropper (with rubber top—1 ml or 1 cc volume)
- Ice (to cool the liquid soap)
- Hot water or heat source (to warm the liquid soap—you do not have to boil it)

- Transparency (onto which you photocopy a sheet of square centimeter graph paper—this provides a grid for estimating surface area of flows)
- Metric ruler (30 cm)

As students work, circulate and review their hypotheses, data tables, experimental designs, and the records they keep of their observations. The investigation presents many opportunities to discuss the nature of scientific inquiry. For example, students will have to determine what to do about bubbles that appear in the liquid soap (should they pop them or not?), how to measure the area covered by the flows, when to stop measuring, and so on. Students are asked to record their results in a data table. They are not specifically asked to graph their results. As you review student work, ask them whether or not a graph of the data might help to reveal patterns in the data. You can prepare an Excel spreadsheet ahead of time, plotting the results obtained by each group for steps 1–7 in such a way that a graph of each group's results is made. This often reveals anomalous data and provides an opportunity to discuss how to resolve and/or deal with discrepancies, or whether or not to average the data to produce a "class average result."

Some students may need help in calculating or determining area and in remembering that area is in units of length or distance squared, whereas volume is in units of length or distance cubed. The investigation provides an excellent opportunity to conduct formative performance assessments of science process skills.

This section provides a link to technology applications. Have students use graphing calculators or spreadsheet software to plot lahar travel times. Alternatively, have them develop their graphing skills with graph paper.

- Graph paper
- Calculator (or computer with spreadsheet program to enter, plot, and print data and graphs)

Opener

Material that erupts from a volcano and flows down its slope is a major concern for people who live near volcanoes. Suppose you live near a volcano and you have just been told that it has erupted.

- What do you think would affect the amount of time you and your family have to evacuate to escape from a volcanic flow? What might control the speed of the flow?

What do you think? Record your ideas about this question in your notebook. Be prepared to discuss your responses with your small group and the class.

INVESTIGATION

Part A: Area of Lava Flows

1. Suppose a volcano produces twice the amount of lava than in a prior eruption. Write a hypothesis based upon the following question: what is the relationship between the volume of an eruption and the size of the area it covers?

   A. Record your hypothesis in your notebook.

2. Check your hypothesis to see if it could be disproved. A hypothesis must be a prediction that can be falsified. The statement "Some stars will never be discovered" cannot be disproved. Therefore, it is not a hypothesis.

3. In this investigation, you will use liquid soap to simulate flow during a volcanic eruption. Volcanic flows include lava, gases, and mixtures of solid particles and gases.

   A. In your notebook, set up a data table. The table should help you record the relationship between volume of liquid soap and the surface area that the soap covers. You will do trials with 0.5, 1, 2, 4, 8, and 16 cm^3 (cubic centimeters) of liquid soap.

4. Place an overhead transparency of a square grid on a flat surface.

5. Pour 0.5 cm^3 of liquid soap onto the transparent graph paper.

6. When the soap stops flowing, measure the area of the flow.

   A. Record the area of the flow in your data table.

7. Wipe the surface clean. Repeat the trials using 1, 2, 4, 8, and 16 cm^3 of liquid soap.

   A. Record your data in your table. Look for patterns.

8. Develop a hypothesis and design a test for one of the following questions related to the flow of fluids. Remember that during scientific inquiry, you can return to the materials or your data and revise your procedures as needed.

   • What effect does temperature have on resistance to flow (viscosity)?

   • What happens to fluid when slope changes from steep to gentle?

   • What effects would you see if fluids moved through narrow channels?

   A. Write down your hypothesis.

   B. Record your procedure in your notebook.

   C. Describe the variables you investigated.

9. Present your procedure to your teacher for approval. Then run your test.

   A. Record your data.

   B. Summarize your conclusions.

   C. Was your hypothesis correct?

10. Obtain results from groups in your class that investigated other questions.

   A. Record their conclusions in your notebook.

Part B: Travel Time of Lahars

1. Examine the table of expected travel times of lahars (mudflows) triggered by a large eruption of Mount St. Helens (see Table 51.1). The values in the table come from computer simulations and actual behavior of mudflows in the 1980 eruption.

2. Make a graph of travel time (in minutes on the vertical axis) versus distance (in kilometers on the horizontal axis) for both data sets.

   A. Plot both data sets on the same graph.

   B. Connect the data points so that you can compare the data.

   C. Calculate an average velocity for mudflows along each fork of the Toutle River.

3. Answer the following questions in your notebook.

   A. Which area (North Fork or South Fork) is more likely to have a steeper gradient? Use the results of investigations from Part A to support your answer.

   B. What evidence in your graphs suggests that the gradient on either river is not constant? Explain.

   C. Based on the information in the table, explain whether or not you think that a community located 50 km from Mount Saint Helens along either of these river valleys would have time to evacuate in the event of an unexpected massive eruption.

TABLE 51.1. **Expected Travel Times for Lahars Triggered by a Large Eruption of Mount St. Helens (Wolfe and Thomas 1995)**

Distance (via river channels) from Mount St. Helens (km)	Estimated Travel Time (minutes)	
	North Fork Toutle River	South Fork Toutle River, Pine Creek, Muddy River, and Kalama River
10	37	11
20	68	30
30	97	54
40	136	81
50	173	109
60	207	140
70	228	173
80	283	211
90	336	258
100	530	312

REFLECTING ON THE ACTIVITY

In this activity, you found that temperature, volume, channels, and slope affect the flow of liquids. Analyzing data from a computer model, you predicted the flow of volcanic fluids down river valleys near Mount Saint Helens. You can now describe the volcanic hazards associated with various kinds of flows and factors that affect the flows. You can also understand how the flows might affect surrounding communities.

INQUIRING FURTHER

1. Research a famous lava flow. Search the Web for information about the Columbia River Basalt group in the Northwest. Prepare a report to the class about the members of this famous basalt group in relation to largest, longest, thickest, cooling characteristics, effects on ancient topography, and cause.

2. Lava and the biosphere: how have lava flows at Mauna Loa and Kilauea Volcanoes affected Hawaiian communities? How does the lava that enters the Pacific Ocean in Hawaii affect coastal ecosystems? What kinds of organisms develop and thrive at the "black smokers" along midocean ridges? Research the 1783 Laki fissure flow in Iceland. It was 40 km long and covered 500 km². How did it affect vegetation and livestock?

3. Lava and the cryosphere: what happens when lava erupts from an ice- or snow-capped volcano? This is an issue in the Cascade volcanoes. Mount Rainier, which overlooks Seattle, has twenty-seven glaciers. Some insights might be gained from exploring the recent eruption at Grimsvotn in Iceland.

ACKNOWLEDGMENT

The activity described in this chapter was developed and tested in classrooms as part of the EarthComm Curriculum Project, which was funded by the National Science Foundation (grant no. ESI-9452789).

SUGGESTED READINGS

Decker, Robert W., and Barbara B. Decker. 1998. *Volcanoes*. 3rd ed. New York: W. H. Freeman.

Francis, Peter, and Clive Oppenheimer. 2004. *Volcanoes*. New York: Oxford University Press.

Sigurdsson, Haraldur, Bruce Houghton, Hazel Rymer, John Stix, and Steve McNutt, eds. 2000. *Encyclopedia of Volcanoes*. San Diego, CA: Academic Press.

Smith, Michael J., John B. Southard, and Ruta Demery. 2002. *EarthComm: Earth's Dynamic Geosphere*. Armonk, NY: It's About Time.

Smithsonian Institute. N.d. Global Volcanism. www.volcano.si.edu/index.cfm.

Williams, Stanley. 2001. *Surviving Galeras*. Boston: Houghton Mifflin.

Winchester, Simon. 2003. *Krakatoa, the Day the World Exploded: August 27, 1883*. New York: HarperCollins.

Wolfe, Edward W., and Thomas C. Pierson. 1995. Volcanic-Hazard Zonation for Mount St. Helens, Washington, 1995. http://vulcan.wr.usgs.gov/Volcanoes/MSH/Hazards/OFR95-497/OFR95-497_inlined.html.

52

A Matter of Timing: Learning about the Impact of Environmental Changes on Animal Migration

Susan A. Kirch and Michele Amoroso

Humans occupy a unique position on this planet because we play several roles. First, as Edward Wilson reminds us in *The Future of Life* (2002), we are organisms connected to and dependent upon the Earth for our lives, just like every other living creature occupying the "thin membrane" that surrounds this otherwise inhospitable rock. Scientists call this thin membrane the *biosphere.* Although many organisms can have a global impact on the delicate equilibrium they have imposed on the physical environment, humans are the only ones who can be conscious and responsible stewards of this fragile planet. Every decision that humans make has ramifications for the living conditions of all life. This lesson plan features an activity on insect migration that introduces students to the complexity of the natural environment and ways that humans can disrupt its balance.

LEARNING MORE

Typically, before beginning a new science unit, we begin by expanding our knowledge on the content. For this study of migration we consulted several sources, including general biology books and textbooks, research journals, science trade books for students, and a variety of websites (see "Additional Resources" at the end of this chapter). Here is some of the information we learned. Migration is the regular seasonal movement of animals from one place to another, often from a breeding site to a nonbreeding site and back. Many animals migrate to find food, such as blue fish and wildebeest. Others migrate to avoid bad weather, like the monarch butterfly, or to reproduce, such as green sea turtles and frogs. The reasons some animals, like the spiny lobster, migrate still remain unclear. True migratory animals make a two-way trip each year from one place to another and back again. The arctic tern makes the longest migration of any animal, from the Arctic to the Antarctic and back again each year (21,750 miles, roughly the circumference of the Earth). The insect that migrates the longest distance is the desert locust (2,800 miles), but the monarch is not far behind (2,000 miles). Recent technological advances have made it possible to accurately track migratory patterns of several different kinds of animals, and scientists have been studying birds, turtles, and whales, among others. Michael Webster and his colleagues (2002) explain that understanding the factors that influence migratory animals throughout their annual cycle is necessary to predict ecological responses to changes in habitat quantity and quality at the locations they seek out and rely on throughout the year.

Scientists are also learning more about how animals migrate to the right place at the right time. Animals measure time with biological clocks controlled by internal sensors sensitive to light. Additional biological meters determine relative position, for example, turtles, songbirds, pigeons, and albatrosses are a few examples of animals that are sensitive to magnetic fields. These animals are able to use their internal sensors to determine a general direction, and it is thought that they use this in combination with other environmental indicators to migrate to a specific address in space. For example, turtles use temperature, salinity, light, chemical, and other sensors to pinpoint the beach where they hatched, and will return to deposit their own fertilized eggs on these nesting beaches. Human impact resulting in environmental changes often disrupts migratory paths and usually has negative consequences. Human activity can disrupt food webs, tax resources, and even lead to species extinction.

BUILDING ON A FOUNDATION OF PRIOR KNOWLEDGE

Before beginning this lesson, we also outlined some general facts and principles we believed our elementary students should be familiar with before we began this classroom inquiry. Given that our students have raised animals and plants in the classroom, we were confident that students were familiar with most of the concepts on our list:

- Students should have had firsthand experience studying living and nonliving objects, and they should understand that living things have properties that are very different from those of nonliving objects.

- Students should understand that most living things seem to breathe, multiply, move, eat, defecate, and respond to stimuli, whereas nonliving things do not possess and control these processes.

- Students should be familiar with life cycles of plants and animals, and understand that stages of the life cycle are often punctuated by a significant event. In the case of butterflies, these events are molting, chrysalis formation, metamorphosis, and emergence.

- Students should know that organisms sometimes have different dietary constraints at different stages of the life cycle. For example, a butterfly larva prefers to eat leaves, while the mature butterfly subsists on nectar provided by specific types of flowering plants.
- Students should be familiar with the concepts *food chain* and *food web*.
- Students should be familiar with climate zones and how they are affected by latitude, altitude, and proximity to the ocean.

With this foundation of basic knowledge, we thought our students were ready to start thinking about more complex interactions between different organisms and between organisms and their environments. This lesson plan suggests one way to introduce students to how human activity can harm a migratory corridor through habitat destruction and, indirectly, through climate changes caused by global warming. We developed a teacher-directed simulation activity ("A Matter of Timing") to engage and introduce students to some possible consequences to disrupting the natural habitats of migratory animals. This was inspired by a chapter (Need Nectar, will Travel: Threats to Migratory Pollinators) in Stephen Buchmann and Gary Paul Nabhan's book, *The Forgotten Pollinators* (1997). After the class analyzed and discussed the implications of the results from the simulation, students developed research questions related to the topic. This shift from teacher-led inquiry to student-led inquiry is essential to provide students with inquiry opportunities in the science classroom.

SETTING THE STAGE FOR INQUIRY

Before beginning the simulation activity, we reminded students that migrating butterflies time their migration to coincide with flowering plant blooms along a corridor. We read Lawrence Pringle's children's book, *An Extraordinary Life* (1997), and asked students to predict what they think would happen if the migration and the bloom were asynchronous.

A Matter of Timing: A Simulation of Flower Blooming and Butterfly Arrival in an Imaginary Migratory Corridor

The students were told that they would be imitating butterfly feeding during migration. Throughout the room, we had set up thirty mock flowers (Figure 52.1) during the prep period. Students ($n = 23$) were told that each cup with red sugar water represented the nectar source of a flower and that the funnel represented the flower bloom. If the large end of the funnel was facing up (Figure 52.1), the flower was considered alive and in bloom. If the neck, or small end of the funnel, was facing up, the flower was "closed," dead, or not in bloom yet. Each student received a tiny cup and a plastic Pasteur pipette. The cup represented the butterfly's capacity for storing energy and was

FIGURE 52.1 Students imitating migrating butterflies are able to drink the flower nectar (*A*, Pasteur pipette) from mock flowers in bloom (*B*, funnel). The nectar reservoir is filled with red-colored water (*C*, narrow-mouthed jar or graduated cylinder).

marked at 5 milliliters (mL) to show that butterfly could only use and store a maximum of 5 mL of nectar. The pipette represented the proboscis. Students were responsible for recording how much nectar they harvested after each round of feeding. To facilitate this, each student was given a data sheet and a clipboard and was assigned to a "data station." Data stations were designated areas to write at around the room.

We decided to conduct ten rounds of feeding representing ten geographic locations along a migratory corridor. Migration maps can be downloaded from the Monarch Watch website (Monarch Watch 2004a, 2004b; a spring migration map is at www.monarchwatch.org/tagmig/spmap.htm, and a fall migration map is at www.monarchwatch.org/tagmig/fallmap.htm). At the beginning of the migration, students closed their eyes while we created a location in the corridor with blooming flowers. For the first location we displayed thirty of thirty flowers in bloom to represent an early, undisturbed location. Students were not allowed to run or shove as they simulated feeding with their proboscis and cups for forty-five seconds. At the end of the feeding cycle, students were instructed to return to their starting positions and discard exactly 1 mL of nectar. Some were dismayed, but we explained that this re-presented the amount of nectar they consumed in order to live, fly, and eat as a butterfly. Furthermore, they were warned that if a butterfly was out of nectar at the end of a feeding cycle, this would indicate that it did not have enough food resources to fly to the next location and was considered dead.

As the students recorded the volume of remaining nectar in their cups on their data charts, we set up the next geographic location along the corridor. We covered fifteen of thirty flowers by turning funnels upside down and capping them with recycled soda bottle caps (Figure 52.2), and reminded students that covered cups represented flowers that were unavailable as a food source.

We repeated this cycle for eight more rounds, mimicking a blooming corridor that was out of synchrony with the migrating butterflies by following with three or four sites that had no available flowers before there was a site with a nectar source. (*Note*: the cycles can be varied to mimic any desired circumstances you would like to discuss. In fact, students were eager to propose alternative scenarios, and it was fun to test them out.)

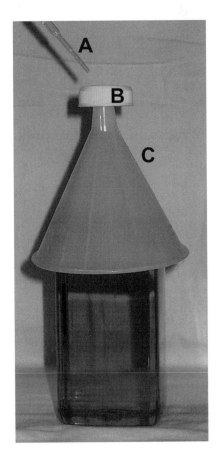

FIGURE 52.2 Students imitating migrating butterflies are unable to drink the flower nectar (A, Pasteur pipette) from mock flowers not in bloom. Mock flowers are represented with a bottle cap (B) resting on the opening of an inverted funnel (C).

ANALYZE THE RESULTS FROM THE SIMULATION

We took many ideas and cues from a practitioner guide edited by Richard Lehrer and Leona Schauble (2002) that describes how to generate and handle data in the classroom. Students graphed the class data, then analyzed and discussed the results from the simulation. Questions we posed during data discussion time included the following:

- What could have caused the asynchrony?
- How do butterflies decide when to migrate and how much time to spend at each location? Is there anything that could change the timing of the migration?
- What determines when a flower blooms? What environmental cues do they sense? What cues could change the timing of the bloom?
- What are the ways you might predict that humans can directly or indirectly impact the nectar available in the migratory corridor?
- Do butterflies and other animals seek out different migratory routes in response to the lack of resources?

DEVELOP RESEARCH QUESTIONS

Students were instructed to write, on a slip of paper, any questions they had at the end of the lesson. We gathered the papers; organized them according to common themes such as butterfly biology, plant biology, migration, climate change, habitat destruction, energy requirements, and food preferences; and posted them for the class. Students interested in similar themes worked together to conduct research using a variety of Internet and print resources to investigate their area of interest.

PARTICIPATE IN SCIENTIFIC INQUIRY: STUDENTS AS NATURALISTS AND CONSERVATIONISTS

In the future we would like our students to become familiar with organisms in their local habitat that are in danger of extinction. We may have the students work with a local researcher and choose or initiate a long-term monitoring project that focuses on local flora and fauna. Alternatively, we have identified several well-established citizen science programs that rely on amateur scientists from across the United States, and sometimes around the world, to submit data about specific organisms, events, or conditions such as Journey North (2005), Citizen Science of Cornell Ornithology Laboratory (Citizen Science 2003), Monarch Watch (2005), and GLOBE (GLOBE Program 2005). National and international projects are exciting for students to participate in, and they provide opportunities for students to gain experience with gathering data, communicating with others, and using those data to answer specific questions. We are looking forward to create additional authentic inquiry opportunities for our students in the future.

REFERENCES

Buchmann, Stephen L., and Gary Paul Nabhan. 1997. *The Forgotten Pollinators.* Washington, DC: Island Press.

Citizen Science. 2003. Cornell Ornithology Lab. http://birds.cornell.edu/.

GLOBE Program. 2005. The University Corporation for Atmospheric Research. www.globe.gov/globe_flash.html.

Journey North. 2005. Annenberg/CPB. www.enchantedlearning.com/coloring/migrate.shtml.

Lehrer, Richard, and Leona Schauble, eds. 2002. *Investigating Real Data in the Classroom.* New York: Teachers College Press.

Monarch Watch. 2004a. Migration & Tagging: Fall Migration Map. www.monarchwatch.org/tagmig/fallmap.htm.

———. 2004b. Migration & Tagging: Spring Migration Map. www.monarchwatch.org/tagmig/spmap.htm.

———. 2005. The University of Kansas Entomology Program. www.monarchwatch.org.

Pringle, Lawrence, and Bob Marstall (illustrator). 1997. *An Extraordinary Life: The Story of a Monarch Butterfly.* New York: Orchard Books.

Webster, Michael S., Peter P. Marra, Susan M. Haig, Staffan Bensch, and Richard T. Holmes. 2002. Links between Worlds: Unraveling Migratory Connectivity. *Trends in Ecology & Evolution* 17 (2): 76–83.

Wilson, Edward O. 2002. *The Future of Life.* New York: Random House.

SUGGESTED READINGS

Attenborough, David. 1995. *The Private Life of Plants: A Natural History of Plant Behaviour.* Princeton, NJ: Princeton University Press.

Campbell, Neil A., Jane B. Reese, Lawrence G. Mitchell, and Martha R. Taylor. 2002. *Biology: Concepts and Connections.* 4th ed. Menlo Park, CA: Benjamin Cummings.

Collard III, Sneed, and Paul Kratter (illustrator). 2003. *Butterfly Count.* New York: Holiday House.

Lasky, Kathryn, and Christoper G. Knight (illustrator). 2002. *Interrupted Journey: Saving Endangered Sea Turtles.* Cambridge, MA: Candlewick Press.

Miller, Debbie S., and Daniel Van Zyle (illustrator). 1997. *Flight of the Golden Plover: The Amazing Migration between Hawaii and Alaska.* Anchorage: Alaska Northwest Books.

Reed-Jones, Carol, and Michael S. Maydak (illustrator). 2002. *Salmon Stream.* Nevada City, CA: Dawn Publications.

Ryden, Hope. 2002. *Wildflowers around the Year.* New York: Clarion Books.

Sayre, April Pulley, and Alix Bereenzy (illustrator). 1999. *Home at Last: A Song of Migration.* New York: Henry Holt.

53

Get Rocks in Your Head!

Michele Amoroso and Susan A. Kirch

Post-Sputnik-era reforms in science curriculum design for early childhood (grades pre-K–2) and elementary (grades 3–6) schools marked the beginning of a national movement for science literacy that included our youngest students. Theoretical frameworks for how people learn, developed by psychologists Jerome Bruner, Jean Piaget, and Robert Gagne, were used by scientists, educational researchers, and sometimes teachers to develop instructional materials for science. The influence of these curriculum developers on early childhood and elementary science is still strong, as seen by the popularity of commercial curriculum kits and teaching manuals (e.g., FOSS, Science Curriculum Improvement Study [SCIS], Delta Modules, GEMS, and AIMS). Recently, however, curriculum selection is being guided primarily by the extent it aligns to state and national science education standards rather than how well it matches current understanding for what we know about how people learn. These two influences are not necessarily at odds, but classroom teachers often find themselves in the position of having a prepackaged curriculum that does not obviously align to local, state, and/or national standards. In this chapter, we show how our class expanded upon lessons outlined in the commercial science kit *Pebbles, Sand, and Silt* (a Full Option Science System, developed by Lawrence Hall of Science; Full Option Science System 2003) to better fit our local standards, and to create inquiry opportunities that ultimately engaged our students in thoughtful discussions about model building and the nature of science.

STUDENTS INVESTIGATE EARTH MATERIALS
AND THINK ABOUT CHANGE OVER TIME

We were surprised to discover that the activities and curriculum structure placed an emphasis on characterizing the size of "earth material" and applying the correct name—for example, cobble, pebble, sand, and silt—to each size group. Teacher-led activities, full of instructions, described how students were to separate a mixture of different sized rock particles using a series of sieves. After sorting a mixture, students were introduced to the correct name for each sized group of rocks—pebbles, sand, and silt—but what was the point? We recognized that this activity would have more meaning for our students if it were done in the context of learning how the Earth changes over time. Words like *pebble, sand,* and *silt* were generic terms that could have been replaced by "rock particles measuring 1–2 cm" or "rock particles measuring 2–4 cm." The terms said nothing about how these different sized rocks were formed, and the activities did not teach students about the process of erosion or its agents of change such as flowing water, chemical reactions, blowing wind, and crushing ice. Furthermore, the lessons did not appear to provide experiences for students to understand that the pebble had once been a part of a boulder and might eventually become a grain of silt if left outside unprotected. We thought students could easily walk away from these initial sorting and characterizing activities thinking that boulders were always boulders and will always be boulders, that sand was always sand and will remain sand, and so on. These statements appear to be true, but are not. Instead, we wanted our students to build a foundation that would help them eventually understand a much bigger idea essential to the rock cycle as described by Art Sussman (2000): "Earth has dry land because the processes that build mountains balance the erosion processes."

Simply by placing the activities in the *Pebbles, Sand, and Silt* (PSS) kit in the context of the rock cycle, we realized we would be able to address two of the three earth and space science concepts that our city required elementary students to understand by fourth-grade commencement:

- Properties of Earth material, such as water and gases; and the properties of rocks and soils, such as texture, color, and ability to retain water
- Changes in Earth and sky, such as changes caused by weathering, volcanism, and earthquakes; and the patterns of movement of objects in the sky

In addition to Earth and space concepts, we also recognized that we could introduce students to unifying concepts in science recommended by the National Research Council in the *National Science Education Standards* (1996), including systems, order, and organization; models, form, and function; change and constancy; and evolution and equilibrium.

With these science-learning goals in mind, we began to think about specific ideas students might find challenging. Our main concern was that the students probably had little experience thinking in terms of geologic time. To

discuss the various ways in which constant erosion and rebuilding are continually changing the earth over geologic time, we decided to read Gerald
Ames and Rose Wyler's (1967) book *The Earth's Story* as a class read-aloud.
We assessed student understanding through expanded measurement and
scale activities used in the math curriculum. After these activities, we believed
the students were beginning to understand that substantial Earth changes
were possible if provided with enough time. Perhaps not surprisingly, students began to ask how old the rocks in their expanding rock collections were
and proposed that maybe they had found the "oldest rock ever!"

After this introduction, we proceeded with the kit activities. The context-
setting work was fruitful. Students seemed to have an easy time proposing
how, over a long time period, a boulder could erode and generate smaller and
smaller particles if exposed to enough wind, rain, and ice. We noticed that the
most common agents of erosion cited by the students were those they had
direct experience with and on a scale they had experience with, rather than
erosion caused by chemical reactions, ice in glaciers, and the flowing water of
rivers and ocean. We were more concerned, however, that students might
think the boulder was gradually disappearing rather than simply disin-
tegrating, so we discussed this idea explicitly and showed video footage of land
erosion by water produced for the Science Curriculum Improvement Study
(SCIS) curriculum kit *Interactions and Systems* (Science Curriculum Improvement Study 1996).

At this point, we were fairly confident that students understood that a
sample of granite or quartz could theoretically be found as a cliff, a boulder, a
cobble, a pebble, a grain of sand, and a particle of silt, and that these terms
were just common names that referred to differently sized particles of Earth
material, but did not identify the mineral composition.

Finally, the readings, classroom discussions, and video footage introduced
the students to the forces of erosion that change the size and shapes of
mountains, boulders, and grains of sand.

The next questions followed naturally: how did the now eroding mountains, boulders, and rocks form? What are the processes that build rock in the
rock cycle? The students were ready to begin the next unit on earthquakes and
volcanism, which they would expand on the following fall in third grade. But
before moving on, we acknowledged that students were bringing in books and
Internet information about Mars and that they were eager to talk about the
types of rocks the Mars rovers, *Spirit* and *Opportunity*, were identifying. Although this was not part of the kit activities or our plan, we thought that
learning about the new Mars discoveries might help us discuss how scientists
work and conduct investigations. This was another opportunity to address
many of the ideas outlined by the National Research Council (1996) under the
"Science as Inquiry" standard, including the following:

- An appreciation of "how we know" what we know in science
- Understanding of the nature of science

- Skills necessary to become independent inquirers about the natural world
- Dispositions to use the skills, abilities, and attitudes associated with science

So we asked if the students wanted to learn more about the Mars discoveries together and build a model of the Martian surface to display at the school science fair. They agreed.

STUDENTS INVESTIGATE
THE WORK OF NASA SCIENTISTS

Through an Internet search for resources on Mars, we discovered that the Jet Propulsion Laboratory of NASA has a website devoted to the Mars Exploration Program (Viotti 2005) that provides clear explanations for educators, parents, and students about what scientists are learning through studying the surface of Mars. The website is filled with stories and news pertaining to the unraveling of the history of Martian rocks and the compelling evidence of the presence of water and its effects on the rocks. Many of the observations that rover scientists are making about Martian rocks are very similar to what the students were observing about their own rock collections, including color, shape, texture, hardness, and size.

To guide student exploration of the Mars Exploration Program website, we asked students what information they believed would be helpful in developing their model Martian landscape. We charted the student questions and statements about Mars, and posted these lists in the school computer laboratory where they would be conducting their informational research projects. During that initial conversation, we noticed that students did not ask about the processes that shaped the surface of Mars, so in the computer lab, we directed them to read through two short stories written and created by Sue Kientz at the Jet Propulsion Laboratory, "The Story of a Little Rock on Mars" (Kientz 1997a) and "An Update from Little Rock on Mars" (Kientz 1997b). These stories provided an entertaining way to make connections to the work students had done learning about change over time due to erosion during the PSS activities.

While browsing the website, students viewed photographs taken by the rover and made several observations that influenced the design of their Martian landscape model. They wanted to show that the surface is a reddish-brown color, that it looks dry (like cracked clay) and rocky, and that the rocks come in many shapes, colors, textures, and sizes. Using earth materials from the playground and the PSS kit, students worked in shifts to build a class model of the landscape. As students built the model of a Martian landscape, we wanted them to understand that models are tentative structures that correspond to real objects and have explanatory power (National Research Council 1996). We asked them to imagine the ideas their schoolmates might have about Mars after viewing the model on display at the science fair. The students did not understand what we meant, so we asked them to think about and respond to statements that a visitor might say. For example, we said, "A schoolmate might think the rocks in your model were actually rocks from Mars and say, 'I didn't know that we could get rocks from Mars!' What would you say to that student about your model?" This example provided the clarification students needed to understand our directive and allowed us to talk about the many similarities and differences between the class' representation of Mars, the model, and the actual surface of the distant planet. This conversation was dominated by one main question from teacher and student: "How do we (or the NASA scientists) know that?"

MISSION ACCOMPLISHED

Today, our students continue to argue about new questions such as whether Mars has an atmosphere, what the temperature is on Mars, whether the ice cap is made of water, and whether the dark areas in satellite photos are large mountains or deep valleys. When they turn their gaze to us for answers, we shrug and ask, "What evidence would we need to answer that question?" or "Hmm, how can we find out?"

REFERENCES

Ames, Gerald, and Rose Wyler. 1967. *The Earth's Story*. New York: American Museum of Natural History.

Full Option Science System. 2003. *Pebbles, Sand, and Silt*. Berkeley, CA: Lawrence Hall of Science.

Kientz, Sue. 1997a. The Story of a Little Rock on Mars. July. www.suekientz.com/little_rock/.

———. 1997b. An Update from Little Rock on Mars. July. www.suekientz.com/little_rock/update.html.

National Research Council. 1996. *National Science Education Standards*. Washington, DC: National Academy Press.

Science Curriculum Improvement Study (SCIS). 1996. *Interactions and Systems*. Berkeley, CA: Lawrence Hall of Science.

Sussman, Art. 2000. *Dr. Art's Guide to Planet Earth: For Earthlings Ages 12 to 120*. White River Junction, VT: Chelsea Green Publishing.

Viotti, Michelle, curator. 2005. Mars Exploration Program. May. http://mars.jpl.nasa.gov.

ADDITIONAL RESOURCES

American Museum of Natural History. 1995. Our Dynamic Planet. www.amnh.org/education/resources/rfl/web/earthmag/index.htm.

Baylor, Byrd, and Peter Parnall (illustrator). 1974. *Everybody Needs a Rock*. New York: Scribner.

Beacom, David. 2005. National Science Teachers Association. www.nsta.org.

Beattie, Laura C. 1994. *Discover Rocks and Minerals: A Carnegie Activity Book.* Pittsburgh, PA: Carnegie Museum of Natural History.

Cole, Joanna, and Bruce Degan (illustrator). 1987. *The Magic School Bus: Inside the Earth.* New York: Scholastic.

Pellant, Chris. 1992. *Rocks and Minerals.* Boston: Dorling Kindersley.

Ricciuti, Edward R. 1998. *National Audubon Society First Field Guide: Rocks and Minerals.* New York: Scholastic.

Windows to the Universe Team. 2003. University Corporation for Atmospheric Research and Regents of the University of Michigan. November. www.windows.ucar.edu.

Wozniak, Carl, webmaster. 2005. National Earth Science Teachers Association. fireton@agu.org.

54

Use of the Internet in the Teaching of Elementary Science

Judith A. McGonigal and Katy Roussos

The Internet Science and Technology Fair (ISTF), developed by the College of Engineering and Computer Science (CECS) at the University of Central Florida, is a model project of how to engage students in authentic inquiry-based research. The competition, at the elementary grade level, invites teams of students from the third through fifth grades to develop a science/engineering solution to a real-world problem in their community. What makes this project unique is that the ISTF competition is entirely web based. To identify, study, and address the real-world problem, students use both the informational resources and social network of the Internet.

To encourage elementary teachers and parents to adopt this innovative model, we share with you our story of how eight fifth graders, in an after-school club, developed an award-winning website for the ISTF competition.

THE COMPETITION

Information, applications, requirements, and reports about the competition are available on the ISTF website (http://istf.ucf.edu). Participants in the competition are required to use the Internet to locate data that help them identify, understand, and solve a community problem. Professionals in industry, government, and research labs volunteer to serve as team advisors who share their expertise with student teams via e-mail mentoring. Students create web pages that present their research and proposed solutions as the final product submitted to judges for the competition.

The highly structured project guidelines of the ISTF competition establish parameters that require participating teams, working as innovative problem solvers, to

- review the National Critical Technology Categories (NCTC) to narrow their project focus,
- meet deadlines,
- submit progress reports, and
- adhere to content and format specifications for the presentation of findings.

BUILDING A TEAM

Initially, our school-based team at Elizabeth Haddon Elementary School, in Haddonfield, New Jersey, consisted of five girls, three boys, and a fifth-grade classroom teacher. None of us knew how to build a website. We expanded our team to include a special education teacher who maintained our school web page. Once we had an on-site technology advisor on our team, we set out to identify a real-world problem as the focus of our research for the ISTF competition.

IDENTIFICATION OF A PROBLEM

The problem identification process was not completely open-ended. According to ISTF requirements, we needed to locate a problem in our community that was directly related to one of the seven National Critical Technology Categories: Energy, Environmental Quality, Information and Communications, Living Systems, Manufacturing, Materials, and Transportation. These technology categories were identified by RAND (1995) as directly linked to the continual growth of America's economy and national security.

After each team member downloaded and read the descriptions of these technology categories, we identified three problems in our community as possible topics for our ISTF investigation:

1. How to keep the park area around a pond free of litter
2. How to prevent the damage to cement sidewalks caused by tree roots
3. How to solve the parking problem in our downtown shopping district

After a lengthy discussion, balancing the identification of a significant community problem with the need for a specific NCTC link, our team finally decided to focus on the parking problem in our downtown shopping area. We wrote the following "Problem/Research Statement" for our ISTF application:

Our downtown commercial area is plagued with the perception of limited parking which negatively impacts the success of small commercial businesses. Parking on the street in front of stores is quite limited. Off-street parking is frequently available, but out-of-town shoppers to our town and potential users of a high speed rail system, which connects our town with two major urban areas, are often unaware of the availability of the parking spaces.

While reviewing the NCTC, we uncovered a possible engineering solution for our parking problem: the use of sensors and electronic communications. In our application to ISTF, we reported that our plan would use the NCTC subcategory of Intelligent Transportation Systems (RAND 1995), which states,

> Intelligent transportation systems (ITS) utilize advanced computers, sensors, electronics, communications, and other technologies to improve the safety and efficiency of all modes of surface transportation for people and goods.

COLLECTING THE DATA

Once the team selected parking as our focus, we needed to document the existence of the problem. The ISTF guidelines required that information be gathered electronically. Students suggested that we e-mail the police department and the mayor to request copies of their recent studies on local parking issues.

Before the team could begin this data collection, we needed to resolve a conflict between our school's Acceptable Use Policy (AUP) and the requirements for participation in the ISTF competition. The AUP is a set of voluntary guidelines designed to ensure the safe and appropriate use of the Internet in our school district. According to this policy (www.haddonfield.k12.nj.us), elementary students are not allowed to use our school district server to send or receive e-mail. We contacted our school district technology supervisor, who suggested that students create pseudonyms to send communications through the teacher's e-mail account. The pseudonym solution had several benefits: students could send and receive e-mail, all e-mail communications could be previewed by the teacher, and the team was permitted to create an online archive of their e-mails, which was linked to our school website.

Using their pseudonyms, students wrote not only to the mayor and the local police department, but also to the online technical advisor provided to our team by ISTF. Our advisor was a manager at Boeing, an aerospace company in Seattle, Washington. Communicating with this advisor was particularly important in the early stages of our project. Having an expert advisor built students' confidence to continue in the competition. Our advisor encouraged us to seek information about a Geographical Information System (GIS) that manages geographical data and a Geographical Positioning System (GPS) that we could use to track people looking for parking spaces. He also helped students consider careers in technology by providing a copy of his résumé and helping the team create a web page that explained the professional roles of scientists and engineers.

LEARNING TO USE NEW TECHNOLOGY

This project brought several types of digital technology out of storage boxes at our school and into the hands of the fifth graders. As the project created a real need for specific skills, the entire team learned how to use new technologies.

To the team's delight, the mayor hand-delivered the police reports and parking studies we had requested. To make use of these data, we needed to find—and learn how to use—a document scanner. With the assistance of the computer lab assistant, the team members learned how to scan printed text and maps, turning them into electronic information that could be posted on our web pages. Later, when the team wanted photographs to provide additional documentation of the parking problems, team members located and learned how to use the school digital camera.

The digital camera also created a new dilemma, which required a creative solution to two sets of opposing requirements. The ISTF content parameters required that information about the team members be posted on the website. Our team wanted to post photographs of themselves. However, our school district's AUP did not allow the faces of students to be shown on any web page created as a school-district project. Although our school technical advisor suggested that the faces in the photographs be blurred, the students decided to manipulate the digital images and cover each face with an animal mask. This solution not only taught students how to enhance digital photographs but also taught students how easy it is to manipulate photographs before displaying them on the Web. Students realized how important it was to view web images with a critical eye. Thus, students gained both a technical skill and technological savvy.

SEARCHING THE INTERNET FOR SOLUTIONS

Once the team had identified and documented the parking problem, the ISTF guidelines required students to electronically locate at least two communities with a similar problem, and report how those communities had solved their parking issue. This task required that team members learn how to use a Boolean operator, which is a form of logic used by most Internet search engines. For example, students searched for *New Jersey* and *parking problems*.

The team discovered many different solutions for parking problems in New Jersey, including the creation of additional parking spaces (Red Bank), the use of parking permits in high-usage areas (New Brunswick), and the design and installation of a robotic device to park cars in a high-rise parking facility (Hoboken). Students found online news articles, as well as government and commercial websites, that described these solutions in more detail. As students compared the various reports, they began to realize that information on the Internet must be evaluated skillfully.

PROPOSING A SOLUTION

After reading the various solutions used in other communities in New Jersey, our team brainstormed various technology solutions for our parking problem. We finally cowrote the following "Project Solution Statement":

Using the sub-category Intelligent Transportation Systems of the National Critical Technologies, we will explore the development and use of sensors to communicate with drivers through computerized bulletin boards, cell phones, and a computerized navigation system about available parking spaces in our town's business district. We believe the development of this technology could be used in many other locations, including solving our parking problem at the two new professional sports stadiums in nearby Philadelphia.

USING E-MAIL TO COMMUNICATE AND LEARN FROM ACADEMIC ADVISORS

Once we had uncovered a possible solution, we spent several club sessions considering how to communicate the specifics of our multiprong solution on a web page. We expanded our team to include the following additional mentors:

- A research engineer at Stanford University
- A research and development manager at Rodel Incorporated
- An assistant professor of corporate finance at the College of Staten Island
- A medical researcher at the University of Pennsylvania

At the end of each session, we communicated our ideas to our technical and academic advisors. During the two-month development of our web page report, these advisors provided suggestions about both content and presentation. Their feedback included spelling and editing corrections, as well as science and engineering information and clarifications. Our advisors also suggested additional experts for us to contact, to gain specific information about computerized display boards, motion sensors, navigational systems, and cell phones. Our team was consistently reminded by our advisors of the importance of precision and accuracy in the presentation of findings. With this electronic mentoring, our team experienced the power of networking with experts to find the best resources and solutions for problems.

MEETING ADVISORS TO ANALYZE EFFECTIVENESS OF SOLUTIONS

In February 2004, when the competition deadline was fast approaching, our advisor from Rodel agreed to come and work at our side at the computer. This advisor taught us how to search patent applications on the Internet. We were amazed to find that ideas similar to our proposed solution had been submitted as patent applications, as recently as one month prior to our Internet search for patents. This advisor also helped us develop a cause-and-effect analysis chart for decision making. Using this tool, we were able to demonstrate why our solution was more effective than even the recent patent applications. Our next step, had time permitted, would have been to develop a cost-benefit analysis. Two advisors reminded us that we needed to consider not only the effectiveness of our solution, but also the cost.

LEARNING TO CONSTRUCT WEB PAGES

The construction of our website was an ongoing process. As we collected and synthesized information, we published our findings on our own website. We utilized Verizon's site builder tool, an application that offered multiple web page templates, each with options to customize by adding photos, text, and graphics. This tool was especially easy for elementary students to learn to use. No knowledge of programming or Hypertext Markup Language (HTML) was needed.

Our team members worked individually and in pairs to develop each web page. We selected the templates that were most appropriate for the content, and then customized the template to reflect personal preferences. We pasted text, uploaded photos and graphics, and developed our own charts and graphic organizers to present our findings. We included hyperlinks to other resources on the Internet. A new page was developed for each component required by ISTF. We linked the pages and organized the website to make navigation simple.

USING A SMART BOARD TO SHARE OUR WEB PAGES

After our web pages were designed and submitted to the ISTF, our team decided to share our solutions for the parking problem to stakeholders in our town. To prepare for this presentation, we learned how to use a SMART board, an interactive whiteboard that can showcase web pages. We set a time to present our research at a Board of Education meeting, to which we invited members of the police department and town council.

CONCLUSION

Participation in the Internet Science and Technology Fair competition provided our team with authentic reasons to

- search the Internet's resources for specific information;
- use the Internet to communicate with advisors around the nation;
- engage in a problem-solving process used by scientists and engineers;

- expand knowledge of how to use technological tools, including a digital camera, scanner, web page builder, and SMART board;
- understand the need to evaluate critically sources and information found on the Internet;
- explore the ways science and engineering interface; and
- engage in scientific discourse with peers, teachers, and advisors.

The Internet Science and Technology Fair models for teachers one way that the Internet can be used in an elementary classroom. This kind of project can integrate science, technology, engineering, and research in the curriculum, while providing multiple opportunities to engage students in critical thinking, collaborative problem solving, networking with experts, and open-ended self-selected inquiry. ISTF (which was renamed Innovations Science and Technology Fair in 2005) provides elementary teachers with a structured set of tasks that encourage students to use the Internet to develop innovative science/engineering solutions for real-world problems.

REFERENCES

Haddonfield School District. 2003. Student Acceptable Use Policy. October. www .haddonfield.k12.nj.us.

RAND. 1995. *National Critical Technologies Report*. http://clinton1.nara.gov/White_House/EOP/OSTP/CTIformatted/.

UCF/CECS. 2004. Internet Science and Technology Fair. September. http://istf.ucf.edu.

ADDITIONAL RESOURCES

Suggested Reading

Lewin, Larry. 2001. *Using the Internet to Strengthen Curriculum*. Alexandria, VA: Association for Supervision & Curriculum.

Websites

Barker, Joe. 2005. UC Berkeley Library: Finding Information on the Internet: A Tutorial. Provides a set of lessons about how critically to evaluate materials found on the Internet: www.lib.berkeley.edu/TeachingLib/Guides/Internet/Evaluate.html

Center for Innovation in Engineering and Science Education. 2003. K–12 Education Curriculum. Provides a catalogue of projects that utilize real-time data or mentors available from the Internet: www/ciese.org/currichome.html

Massachusetts Department of Education. 2001. Selected Websites for Science and Technology/Engineering Education. Provides links to web sites that help K–12 classroom teachers integrate science, technology, and engineering into their curriculum: www.doe.mass.edu/frameworks/scitech/2001/resources/web.html

55

Using *Bill Nye, the Science Guy* within a Framework of Inquiry Science

James Truby

THE TECHNOLOGICAL CLASSROOM

When I was a first-year science teacher in 1967, I would look out at my class and think, "I wonder if anyone understood what I just taught." I fantasized, "Wouldn't it be wonderful if each student had a magic button that they could press that would register their answer to any question I asked?" I would have immediate feedback and know if they understood. That magic button is now a reality in my classroom, and teaching will never be the same. The magic buttons are the ACTIVoters that are used with Promethean's ACTIVboards. Together they have changed the face of classrooms and what goes on in them. Here are some examples of how I use my ACTIVboard and ACTIVoters.

As a sixth-grade student enters my room, he reads the tickertape scrolling across the interactive board. It reads, "Today we will perform the Fantastic Flames Lab." Behind it, they see an Astronomy Picture of the Day. Today it is a Hubble photo of the youngest galaxy in the universe. Today's students know what it is like to sit in a movie theater; my classroom is sort of an interactive, challenging, science movie theater.

When doing the discussion of the laboratory investigation, the students periodically enter answers to questions I pose, using their electronic voters that they nickname "turtles." The percentage of correct answers is displayed on the interactive board, and the teacher knows whether the idea has been mastered, as do the kids, or if a review is needed. If so, the advantage seems to be a democratic, shared, learning process. "Oh nuts, we all missed this one. Let's try it again!"

The technology also acts as a radical time saver for me. On the day of the test, questions appear on the board. There are pictures of Hubble's deep field and accompanying questions. There are graphs to interpret and pictures of the spectrums of receding galaxies to analyze. When the test is finished, I hit "Save As," and the test grades are stored on a computer.

At night, parents go online to check the latest posted grades. There is an accompanying newsletter that explains what each graded activity means and all that was expected of the student to achieve that grade. For example, there is a clear description of what the test covered or what was done during a lab. Some parents decide to e-mail me to comment on their child's progress or perhaps to make arrangements for work to be made up.

A young lady who was absent for an exam comes before school, and we enter her ACTIVote code number. We do this so her grade will be stored in the computer under her name. After she uses the electronic pen to bring up a question, she uses the ACTIVoter to answer it. When she is finished, she hits "Summary," and a graph appears showing how she did on each question and her total score. I hit "Save As," and the grade is recorded.

A student comes up before class and says, "I was absent yesterday; what did I miss?" I answer, "We did the lab conclusion," then I go to the ACTIVboard and open the class folder for the missed hour. The ACTIVboard displays a sequence of flipcharts of all the missed work. I hit "Print," and within seconds there is a copy of the data, pictures, charts, graphs, and conclusions of the previous day's work.

My technological classroom actually came about as a result of both my wonderings and something that was said to me five years previously by Ken Tobin. As is his habit, he was blunt and succinct. When referring to the work I had done with his student intern, he said, "In the future if you have an intern of mine, I would like them exposed to the use of computer technology in the classroom." Several years later when we met, I was expounding the virtues of my technological classroom as I wanted him to know I had paid attention to his previous request. But his comment had nothing to do with the technology. He said something more startling: "I hope you have not lost the common touch with your students." I was stopped dead in my tracks. The reality is that using technology intelligently can augment the common touch. What I have learned since is that *technology* is not synonymous with *impersonal*. Solving student problems before they become serious through parent e-mails, answering student e-mails that ask for help in completing hyperlinked homework assignments, e-mailing students to say that it is okay for them to come in before school to work on an experiment, bringing up websites on the interactive board, having all previous work stored on a computer to share with students, watching students use their "turtles" to make up tests, and seeing their joy as the screen announces that they have successfully answered another question correctly increase one's personal touch with students. Its very efficiency frees the teacher to spend time with the student in a personal way.

LEARNING TO LOVE SCIENCE

During the 1970s, I taught biology at Bishop Moore High School in Orlando, Florida. Dr. Herbert Hellwege, the chairman of chemistry at Rollins College in nearby Winter Park, asked if he could sit in on my classes. I told him I would be honored, and after that, occasionally, I would notice him sitting in the back of my classroom. I never asked him why he came. One day he stopped me in the parking lot and said, "Aren't you the least bit curious about why I come to your classroom?" He told me that an inordinate number of students from our high school signed up for his chemistry class, and when he inquired why, they told him that they loved their high school biology class with Mr. Truby and for them science had become fun. The professor said he wanted to see what all the fuss was about. Then he said, "You have made them love science." He continued,

> As you might know, the Busch foundation has recently granted our college the money for a new science building. We have fantastic facilities equipped with the latest of everything. However, if the students do not love science, I will never see them in my classes. Please, keep on doing what you are doing. I have the staff and facilities to teach them everything they need to know about science, but you make them love it.

I was reminded of this conversation recently when, during a school science meeting, we were told to be sure to teach all the curriculum guidelines so our students would perform well on the state-standardized science test. We were further told that part of our motive could be the prestige of being rated an A school, and a $1,000 bonus for each teacher. I wondered silently what this had to do with the excitement of science education and what it had to do with teaching science so that students would love it and continue with it at a higher level. My answer: "Not much."

Loving science has implications far beyond our classroom. For three years during the 1990s, Dr. E-ni Foo of the University of Pennsylvania flew anywhere from 50 to 100 teachers from the Far East who were participating in the Benjamin Franklin Institute to Orlando so they could visit my classroom at Rock Lake Middle School. I loved having them in my class. During a roundtable, they troubled over the fact that although their students led the world in science and math testing, they weren't producing scientists. They wanted to see how we did it. For me the answer was that our students, my students, learn to love science.

The rule of thumb I use in deciding how I will teach something is simple: will it be fun? After that I find the most exciting laboratory investigation to teach that something. Often that activity will start with what, at first, looks like no more than a "hook." That hook is designed to be so exciting that they will clamber to learn anything associated with it. Here is an example:

Perhaps you want your students to experience what Edwin Hubble did when he discovered receding galaxies. Most sixth-grade classes do not have the equipment to photograph the spectrums of receding galaxies. But you can flame test metal salts in most classrooms, which leads to examining the flame tests through spectroscopes, which leads to going online to see what spectrums look like when galaxies move toward or away from us, and finally to the work of Hubble and red shifting. Along the way, when students want to know where the colors come from or why the colors are different, they can learn to build atoms, putting the electrons in the proper orbitals. Then they can learn about the electrons becoming excited. This leads to neon lights, the auroras, and emission nebulas.

This is a hook.

Of course, you could do it the old way: have them draw dots on a balloon and blow it up as Sir Arthur Eddington did: "See—the universe expands." But this is not as exciting. Actually, I will admit that my students do stick dots on balloons, inflate, and watch the dots move apart. We discuss what the inside and the outside of the balloon represents. It is still a darned good activity.

USING BILL NYE IN THE SCIENCE CLASSROOM

While a great deal of knowledge is generated as a result of exciting investigations, there is a wonderful existing resource for presenting knowledge to students. Every Friday after I pass out the snacks and the students have retrieved their drinks from the refrigerator, we turn out the lights. I hit the "Play" button on the VCR, our interactive board is transformed into a movie screen, and our room is transformed into a movie theater. Within seconds, the kids are singing, "Bill Nye the Science Guy." Just a spoonful of sugar makes the medicine go down. On any particular Friday, they could be watching "Chemical Reactions," "Light," or "Measurements." On this particular Friday, they are watching "Atoms." When the lights come back on, they become aware of the balloons that they have hung from the ceiling. In anticipation of this day, they have been building atoms. In order to understand the distribution of electrons in the atom's orbital, they have used balloons of many sizes, shapes, and colors. There is a little red balloon blown up inside a large white balloon, the 1s and 2s orbitals. There are three wiggly balloons blown up and twisted in 3-D to look like 2p orbitals. They have used the interactive board to build atoms and know the electron distribution of at least the first ten elements on the periodic table. They recall Bill telling them minutes before, "The Periodic Table, never leave home without one."

There are tons of things to discuss. After I mop up the spilled drinks and sweep up the crumbs, we proceed. How far did Bill run from the nucleus represented by a vibrating dog toy to the nearest electron? Answer: 500 meters. The atom is mostly empty space. Why can something so empty feel so hard, such as the table they are resting their elbows on? They recall Bill trying to stick a piece of paper through a running fan.

Bill Nye has packed his presentations with humor. One of the recurring characters is Richie, who is known to the students as the boy who eats the centers out of Wonder Bread but never eats the crusts. Richie's father has experimented with his famous brownies recipe and leaves the "O" out. He gets a hydrocarbon $(H + C)$ rather than a carbohydrate $(H + C + O)$. The brownies taste like gasoline. Bill has explained organic chemistry as he dodges cow pies and an irate bull in a cow pasture.

Some of the humor goes right over their heads, like the Atomic Saloon, a parody of the 1982 movie, *Atomic Café*. Some rough-and-tough cowboys are playing poker with element cards in the Atomic Saloon. "Can anyone beat $C_6H_{12}O_6$?" one dude asks. "Yup, $C_{10}H_{14}O_7$," another boasts in a thick Texas drawl. They look at each other and yell in unison "Oil of spearmint," and a brawl ensues when it is discovered one cowboy has a hydrogen up his sleeve.

Bill always presents experiments that the students can perform at home for fun, extra credit, or both. In this episode there is molecule packing with water and alcohol. When they watched "Heat," Bill told them how to make a hot fudge sundae from scratch. Afterwards many students wanted to make sundaes at home. We drew pictures of how to make one on the ACTIVboard, and after hitting "Print," I instantly had copies to give them.

The first week of school, my students are exposed to Bill's healthy skepticism. They hear, "Extraordinary claims require extraordinary proof," as they watch "Science vs. Hoax." Scientists try to understand the world by finding these extraordinary proofs. For example, "The world is round" is introduced as an extraordinary claim. The extraordinary proof proceeds like this: the shadow cast on the moon by the Earth is always round, and Bill's ship model of science disappears down the horizon, tip of the mast last.

If you can test it, it is science. The students are asked, "Are pictures extraordinary proof?" as they are shown a photograph of the Lock Ness monster. Bill leads a discussion while in the background, the faked picture seems to come alive as a diver with a Lock Ness monster glued to his head is attacked by the cameraman who has been tricked into thinking the monster is real. The kids see that as convincing as the diver-monster might be, he is a hoax—not science.

Bill Nye's real science and a deep dose of humor are terrific teaching. Bill is the one-ton man in the "Science vs. Hoax" episode. I have shown this episode for years, and every year I anticipate him leaning on the balsa wood table or sitting in the balsa chair. The set shakes with his every step. I laugh every time I see it. Extraordinary proof? I do not think so. The kids learn you have to be able to weigh him on a variety of scales; you have to repeat your tests.

In the same episode, James Randy, a world-renowned magician, demonstrates his illusion skills in bending spoons between his fingers as he asks the students "If I can do it this easily, what do you think of people who claim they do it with psychic power?"

Later in the episode the students are instructed how to order their Hoaxster 2000 Kit, complete with fake Bigfoot feet, a flying saucer, an alien mask, and an affidavit saying that they have been abducted by aliens. Bill picks up on the alien theme and asks, "Do you think there are flying saucers? Extraordinary claims require extraordinary proof." He explains that the Clark Reflector has been exploring billions of galaxies for 100 years. Have they found one iota of proof of flying saucers? Nada—zip!

Each episode of *Bill Nye* instills the spirit of science into students. Of course the show becomes dated as the body of scientific knowledge expands and is replaced with new knowledge. He begins his "Outer Space" with the question "How long would a tape measure have to be to measure the universe?" Bill says, "Ten billion light years." For students who had just seen the Wilkinson Microwave Anisotropy Probe (WMAP) on the Astronomy Picture of the Day, this is 3.7 billion light years off. This is great for them to see; after all, students need to see that as our scientific body of knowledge grows old, scientific truths disintegrate. And so we ask the students to question everything. Another example of this occurs in Dr. Timothy Ferris's "Creation of the Universe," which my students viewed at the beginning of our unit. This is a marvelous introduction to the cosmos created by the well-known UCLA professor, Dr. Timothy Ferris. He shows a time cone and takes the viewer back 5 billion years to when the galaxies are no longer found. One of my students, who had checked out the Astronomy Picture of the Day on her birthday, as part of an assignment, came up with a picture of a galaxy 13 billion years old. This is not going to stop me from loving Ferris's introduction to the cosmos, nor will I stop showing Bill Nye's "Outer Space," because his tape measure is 3.7 billion light years too short.

Perhaps you want to sample Bill Nye before you use him. I invite you to question me as we invite our students to question everything. So get Bill Nye at his best in "Comets." Hear Half Pint say, "Hey Pa, it landed on our prairie, can we keep it?" on "Little Meteorite on the Prairie." Enjoy Meteor as he says, "Gee Wally, if I name it a meteorite the kids will razz me and call me a goof." On "Leave It to Meteor." Thrill at the latest toy craze, "Tickle Me Meteorite." Find out what happens when you pull his nose. Finally, learn about ellipses on "Alfred Hitchcock, Make That, Bill Nye Presents." This is great stuff.

COMMUNICATION

The most important communication skills that I have learned, I learned from Fred Rogers. Unlike other television hosts, who feel they must fill every silence with their talk, Fred Rogers listened. When I was teaching high school during the 1960s and 1970s, I made it a point to watch *Mister Rogers' Neighborhood*. I was intrigued that when he had a guest, the guest did most of the talking and Fred made the guest the center of attention. He asked questions that he genuinely wanted to know the answers to, and then would carefully listen. It was obvious that his listening led to other genuine questions. Even when he was talking to his TV audience, he would ask questions and make them know that they were a part of what was happening.

Rachael Remin, in her book *My Grandfather's Blessings*, explains that the Dalai Lama possesses the same trait. She tells the story of her friend standing in a room filled with important dignitaries. The Dalai Lama entered via the back door. He immediately began talking to her. The dignitaries in the front of the room seemed to melt away, and she felt that only she existed. This is the kind of communicating we as teachers need to cultivate.

For thirty-eight years I have been involved in inquiry teaching in which what is learned in the classroom grows out of open-ended laboratory investigations. That is to say that what the student learns as an outcome of the investigation is not predetermined. Perhaps the most important times in inquiry teaching is when the students report what they have observed. It is important that they feel what they have seen is important. No observation is unimportant. I must be constantly alert, often repeating what has been said to be certain I understand what the student is saying. I have to ask intelligent questions to clarify what I have heard. During one of their visits, the teachers from the Far East stated their surprise that I spent so much time listening to my students and using what they had said as the basis for my lessons. They were used to dispensing knowledge. They were also surprised that my students felt a total lack of intimidation and possessed the freedom to challenge me or what was being said in class. This was very foreign to them. I pointed out that such open inquiry is essential in science. In inquiry teaching, students do what scientists do: they challenge everything.

TEACHER AFFIRMATION: A CONCLUSION

How do we know if what we teach is valid? Do we ever receive validation? I do not use a textbook, so I cannot rest assured that experts confirm that what I have decided to teach is right. Even if I had such expert confirmation, I am reminded of Richard Feynman's unpleasant experience serving on a science textbook committee. He relates how he found the quality so abysmally low that he withdrew from the committee. I am also aware of the State of Florida's attempts at defining the science curriculum and its failed attempts with its State Standards of Excellence and the Revised State Standards of Excellence. To be candid, I am a tad skeptical of the "experts."

Affirmation? The incident with Dr. Helwiege at Rollins College remains with me. In a sense, he told me that the medium is more important than the message. To learn to love science is more important at the secondary level than swallowing scientific facts. This same message was delivered when the Far Eastern teachers told me their students lead the world in science test results but did not love science enough to become scientists. That is why the experts from the Far East came to see me teach.

Ten years ago I had my first meeting with Ken Tobin. I remember him slipping into my room before school that first day, catching me working with some students. We were doing spectroscopy work with a Tesla coil and some large handblown glass balls containing various inert gasses. I recall hearing him clicking away at his laptop computer and seeing him dancing in the back of the room saying, "I have waited all my life to see this." The "this" was the nature of the exchanges taking place between my students and me and the energy those exchanges created. Much later I learned that what he saw that day was included in an article that he published.

Several years ago I was put in touch with several top astrophysicists and astronomers, Dr. Nat Carlton at Harvard University and Dr. Craig Foltz at the Smithsonian Multiple Array in Tucson, Arizona. They agreed to answer students' questions about the cosmos that I could not answer. It turned out that those questions came every day. For several years, they fielded questions via e-mails. My students asked and asked . . . and learned. We knew things long before they were in textbooks or, for that matter, before they made television specials. Nat became enamored of my students and told me he wanted to meet the kids who could ask such great questions. Sixth graders do not know enough to be inhibited in questioning; I found myself thinking, "I wish I had thought to ask that." On one of his trips to Florida, where he was building a catamaran to sail around the world, Nat stayed with me and visited my classroom. What a day. They flooded him with questions, not just cosmic question, but questions about how he had become a physicist and what things he liked to do.

Recently during our lunch club, a takeoff of the movie *The Breakfast Club*, when students who do not want to eat in the cafeteria watch science videos in my classroom, we were watching Bill Nye's "100 Greatest Scientific Discoveries." After watching the episode on astronomy, I asked, "What did you notice new about what you just saw?" Almost immediately, they responded that they had been exposed to everything that they had seen. I took it a step farther and asked, "What did he leave out?" This took longer for them to respond, and then several said, "The crazy guy." That was exactly what I was thinking. Bill had mentioned the discovery of gamma ray bursts from supernovas and their importance, but had failed to interview Bohdan Pazynski, the man who figured them out and whose hypothesis had been ridiculed. My students had seen him on a *NOVA* episode entitled "Death Stars." Not only did they know the material, but they also knew what was missing: the guy who was "crazy" enough to ask apparently crazy questions!

Last year I was observed by a team from the county office, as part of an effort to determine the Seminole County teacher of the year. I remember a comment they made to my principal when she was notified that I had been selected as 2005 Middle School Teacher of the Year. They told her when they read my résumé and discovered I was sixty years old, they expected to find a dinosaur in my classroom. Instead, they found a teacher, excited—no, in love—with his subject and filled with questions. Listening to observations and still more questions. I was so proud of my class—just doing what they did every day that year—and loving it.

WEBSITES

Bill Nye, the Science Guy tapes and compact discs can be ordered from Disney Education Productions: www.edustation.com

Information concerning the ACTIVboard and ACTIVoters can be found on the Promethean Inc. website: www.prometheanworld.com

PART 8

SCIENCE IN THE NEWS

56

The Origin of the Chemical Elements

John Robert de Laeter

INTRODUCTION

Every person is fascinated by the night sky. We have all asked ourselves questions such as "How did the universe come into being? What is going to happen to it some time in the future? How were stars like our Sun born, how do they live, and will they die?" Science teachers are confronted by these and other questions about our universe by eager, enquiring minds. Those questions deserve an answer! And what is the origin of the chemical elements on Earth? Where do the oxygen we breathe and the calcium in our bones come from? Why is gold such a rare element, whereas iron is mined in huge quantities?

THE ABUNDANCE DISTRIBUTION OF THE ELEMENTS

What, then, is the distribution of elemental abundances in the universe? The present-day abundance curve is the product of history and was shaped by cosmic events. From this curve we can learn much about the "big bang," the evolution of stars, and the nuclear processes that produced this abundance distribution in the solar system, and by inference in other parts of the universe. It was not until 1925 that astronomers realized that approximately 99 percent of the universe was comprised of H and He, with H being about ten times more abundant than He. However, the first reliable solar system abundance curve was not produced until 1956, as shown in Figure 56.1 (Suess and Urey 1956). Up to about fifty years ago, many scientists

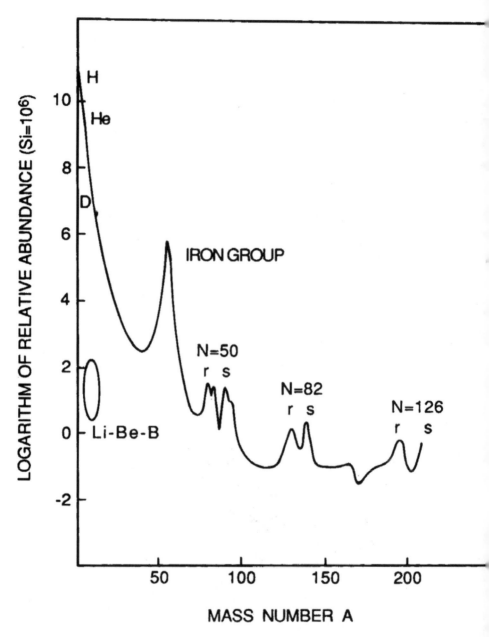

FIGURE 56.1 A schematic curve of nuclidic abundances (relative to $Si = 10^6$ atoms) versu mass number A for the Sun and similar main sequence stars. The symbols s and r stand for th slow and rapid neutron capture processes, respectively.

believed that all the elements were created at the beginning of the universe in that high-temperature, high-density epoch we call the *big bang*, in which an initial mix of elementary particles led to the synthesis of H from which all the other elements were built up by successive neutron captures. Unfortunately no nuclides of mass 5 or mass 8 exist in nature, so it was difficult to see how a step-by-step neutron capture process could synthesize any nuclide with mass number $A > 4$. Not a promising start to synthesizing the ninety-two elements!

In 1957 the nuclear processes that could explain the abundance distribution curve of Hans Suess and Harold Urey were identified by Margaret Burbidge, Geoffrey Burbidge, William Fowler, and Fred Hoyle (1957). They argued that the two elements with the lowest charge number Z—H and He— were in fact derived from the big bang, together with a small proportion of Li ($Z = 3$). However, all the other eighty-nine elements that exist in nature were continuously synthesized in stars. "Creation" was not a single event that occurred some 14×10^9 years ago at the time of the big bang, but is a continuing phenomenon. The seminal paper of Burbidge and her colleagues identified two major aspects of element nucleosynthesis:

(a) Charged particle (or thermonuclear) reactions that are responsible for synthesizing most of the elements up to the iron-peak group of elements
(b) Neutron capture reactions that dominate the production of the "heavy" elements beyond the iron-peak group ($A \geq 65$)

THE LIFE CYCLE OF STARS

Like human beings, stars experience a life cycle in that they are born, they live, and they die—and the duration of their life depends inversely on their mass. Their "deaths" can be orderly as they simply fade away, but massive stars can have spectacular and violent deaths in which their nucleosynthetic products are spewed into the surrounding space by supernovae, thereby providing the material from which new stars can be born. So stars are born in cycles. Our own star, for example, is only 4.5×10^9 years old, and was therefore not in existence at the time of the big bang. It has been formed as a third- or even higher-generation star from the nucleosynthetic debris of other dying stars—which is why we have all ninety-two elements present in the Earth.

A first-generation star is formed early in the history of the universe from a mass of H atoms and smaller amounts of He and Li. This whirling mass of gas contracts under gravitational attraction until sufficient heat is generated by collisions to produce a temperature of approximately 10^6 K when thermonuclear fusion can occur. Thermonuclear fusion, in its simplest form, is the process whereby four protons $_1^1H$ are converted to a nucleus of helium $_2^4He$ by the reaction

$$4_1^1H \rightarrow {}_2^4He + 2e^+ + 2\nu + Energy$$

In astrophysics, we refer to this reaction as *hydrogen burning*. In addition to the single He nucleus, two positrons and two neutrinos are also produced, accompanied by the release of nuclear energy. Sufficient energy is released to enable the dense mass of gas to become luminous—and thereby a star is born!

As hydrogen "burning" continues in a main sequence star such as our own Sun, a core of He develops in the center of the star until a significant fraction of the H in the interior of the star is exhausted. The equilibrium between inward gravitational forces and the outward forces produced by these nuclear fusion reactions is then disturbed, so that the core contracts, causing higher temperatures, which in turn cause the outer envelope of hydrogen to expand. The extended surface radiates at a cooler temperature than was the case for the star in its main sequence phase, and the star is then known as a *red giant*.

What happens next? We have noted that no nuclides exist at masses 5 and 8, so how is He of mass 4 able to produce C and O and heavier elements so vital for life? It is possible, of course, for two He nuclei to collide to produce a Be nucleus of mass 8, but this nucleus is unstable and breaks down again into two He nuclei such that

$$ {}_2^4 He + {}_2^4 He \leftrightarrow {}_4^8 Be $$

However it is possible, under certain nuclear conditions, that a third He nucleus may interact with the unstable Be nucleus before it has a chance to decay, forming an excited state of ${}_6^{12} C$. Thus,

$$ {}_4^8 Be + {}_2^4 He \rightarrow {}_6^{12} C $$

It so happens that this excited state of ${}_6^{12} C$ has a small probability of surviving due to a resonance between the collective energy of the He nuclei and an energy level in the ${}_6^{12} C$ nucleus. This amazing set of conditions, which only occurs at the temperatures and densities found in the core of red giant stars, coupled with the unique energy-level structure of ${}_6^{12} C$, enables C to find its way into the chemistry of life. Fred Hoyle has said that "it almost seems as though the laws of physics have been deliberately designed with regard to the consequences they produce inside stars."

Thus the net effect of helium "burning" is to convert He into C and O and thus circumvent the absence of stable nuclides with mass 5 and 8 by "leap-frogging" over them. The Suess-Urey elemental abundance curve in Figure 56.1 shows that Li, Be, and B are low-abundance elements due to the fact that they are not produced in stellar nucleosynthesis, but rather by cosmic ray interaction with interstellar gas and dust.

When most of the He is exhausted, the core contracts once again, increasing the temperature from approximately 5×10^8 K to 10^9 K when carbon and oxygen "burning" can occur, together with a network of alpha capture and other capture reactions. The higher temperature is required to overcome Coulomb repulsion forces inherent in the heavier nuclides. These nucleo-synthetic reactions continue to produce heavier nuclides until the iron peak of elements is reached, when these thermonuclear reactions become ineffective in producing elements with a mass number greater than that of Fe because no further energy is produced.

THE SYNTHESIS OF THE HEAVY ELEMENTS

The nucleosynthesis of most of the nuclides heavier than the iron group elements are produced by an adaptation of the neutron capture theory. A second-generation star inherits a certain number of elements up to and in-cluding Fe, in addition to H and He. As before, hydrogen "burning" will occur as the star moves up the main sequence. However, when the star reaches the red giant phase, neutrons are produced that enable the heavier elements up to Bi to be synthesized by the slow (s)-neutron capture process. The s-process proceeds on a slow time scale and involves a succession of neutron captures each followed by a series of beta decays until stable nuclei are reached. The capture path for the s-process follows the "valley of beta stability" through a succession of stable isotopes (Figure 56.2). The second neutron capture process (the r-process) occurs on a rapid time scale of a few seconds in duration in explosive events, including supernovae. In a rapid flood of neutrons, unstable neutron-rich nuclides are synthezised, which subsequently beta decay until they achieve stability when the intense neutron flux is removed. The r-process can produce the elements beyond Bi, including the transuranics.

Figure 56.2 shows the s-process path through the isotopes of Cd, In, Sn, Sb, and Te. Tin is particularly interesting because it has more stable isotopes than any other element, a fact that reflects its magic number of protons ($Z = 50$). Masses 116–120 lie on the s-process path, but of these ^{116}Sn is an s-only produced nuclide, since it is shielded from r-process production by the stable isotope ^{116}Cd. Likewise, ^{110}Cd, ^{122}Te, ^{123}Te, and ^{124}Te are also s-only process nuclides. The s-only process nuclides are formed in the red giant phase of stellar evolution over periods of time up to 10^7 years in duration. The product of the solar system abundance, N_S, and the appropriate neutron capture cross-section, σ_{RG}, should be a constant over a restricted mass range (Burbidge et al. 1957). Neutrons with an energy distribution appropriate to those found in red giants can be produced in the laboratory, and thus s-process systematics can be tested experimentally. For the case of Te, the $N_S\sigma_{RG}$ products of the three s-only process nuclides are essentially constant, thus demonstrating the validity of this aspect of heavy element nucleosyn-thesis.

An examination of the heavy element region of the Suess-Urey Solar System abundance curve in Figure 56.1 reveals three distinct double-humped abundance peaks. These are caused by the extra stability associated with

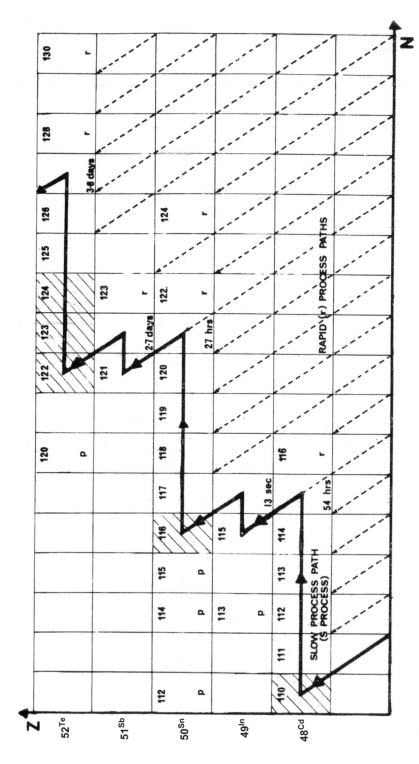

FIGURE 56.2   Chart of the nuclides in the mass region from Cd to Te. The s-process neutron capture path is indicated by the heavy black line showing the s-only process nuclides. Also indicated are the r-process decay chains and the r-only nuclides.

"magic" neutron numbers $N = 50$, 82, and 126. The reason for the double-humped nature of these peaks is because of the existence of the two neutron capture processes that follow different pathways in an N–Z diagram. It follows that nuclides with a large neutron capture cross-sections at astrophysical energies, such as ^{197}Au, which is converted to ^{198}Hg, will have a low abundance, whereas $^{208}_{82}Pb$ is much more abundant because it has a doubly magic $Z = 82$ and $N = 126$. Thus, Au is a rare element in nature because the nuclear furnace in which it was generated also effectively transmutes it into Hg by neutron capture.

CONCLUSIONS

The origin of the chemical elements involves the synthesis of nuclides by nuclear processes and is intricately linked with the evolution of stars and energy generation. It was Rutherford who showed that one element could be converted to another by nuclear transmutations, and this led Arthur Eddington to declare that "what is possible in the Cavendish Laboratory may not be too difficult in the Sun."

The universe comprises 92 elements, 286 stable isotopes, and an enormous number of radioactive isotopes. Each isotope has its own unique nuclear composition, but each isotope possesses the fundamental building blocks of protons and neutrons that can be rearranged into a variety of nuclei. The understanding of the synthesis of the elements was hidden in the abundance distribution of the elements, and its solution is one of the triumphs of twentieth-century science. The relationship between nuclear physics and observational astronomy is the new discipline of nuclear astrophysics, which has enabled us to relate the various nucleosynthetic processes to the evolution of stars from interstellar gas and dust to the main sequence, red giant, nova, or supernova stages, and the functional dependence of the type and time scale of this evolution with the mass of the star.

The poet Walt Whitman was close to the mark when he said, "I believe that a leaf of grass is no less than the journey work of the stars."

The detailed verification of s-process neutron capture by laboratory-based experiments has been most impressive. A study of isotopic anomalies in various now-extinct radionuclides in primitive meteoritic material has provided convincing evidence that a supernova injected fresh nucleosynthetic material into the region from which our solar nebula formed, thus triggering its collapse to form our Sun. There is food for philosophical thought in what we have learned about the origin of the chemical elements. The elements that comprise our own bodies were synthesized in a stellar furnace, and have taken billions of years to produce. Many cosmic events preceded the formation of the Earth, and our solar system was not even in existence at the beginning of our galaxy. So dies the last vestige of mankind's geocentric conception of our role in the universe.

REFERENCES

Burbidge, Margaret, Geoffrey Burbidge, William Fowler, and Fred Hoyle. 1957. Synthesis of the Elements in Stars. *Reviews of Modern Physics* 29:547–650.

Suess, Hans, and Harold Urey. 1956. Abundances of the Elements. *Reviews of Modern Physics* 28:53–74.

ADDITIONAL RESOURCE

Clayton, Don. 1983. *Principles of Stellar Evolution and Nucleosynthesis.* Chicago: University of Chicago Press.

57

The Big Bang: How the Universe Began, and How to Teach It

Harry L. Shipman

INTRODUCTION

From the earliest times, human beings have looked up at the starry sky, contemplated this wonderful world we live in, and asked questions. "Twinkle, twinkle, little star, how I wonder what you are?" Are stars suns like our own? How far away are they? Where, after all, did they come from? (See Figure 57.1.)

Answers to these questions are as old as the questions themselves. Over time, however, the answers slowly slid from the mythological domains into the scientific ones. Why does the Sun rise and set? Having one of many Greek gods carrying the Sun across the sky on a chariot was at one time seen as an adequate answer. But this unverifiable story was, in time, replaced by the rotation of the Earth as an explanation. Astronomy as a science began with the heliocentric model of Copernicus, the exquisite observations of Tycho Brahe, and the genius of Johannes Kepler, Galileo, and Isaac Newton, which combined the Copernican model, laws of physics, and precise observations into a comprehensive explanation of the motion of planets. By the beginning of the twentieth century, we had a pretty good picture of what the Sun was, where the planets were, and what the stars were. But we still didn't know where it all came from.

One hundred years later at the beginning of the twenty-first century, we have a scientific answer to this question of origins. It is supported by evidence, accepted by the scientific community, and has proven to be a very fertile explanation that suggests a whole host of additional questions. It deserves the status of the name *theory*, not just in the popular sense of a reasonable guess,

FIGURE 57.1 **The Hubble Heritage Team combined data from two
different projects in order to produce this picture of the globular
cluster Messier 80. The designation refers to a 200-year-old catalog
in which this object first appeared. Globular star clusters like this
one are some of the oldest structures in the universe.**

Image credit: The Hubble Heritage Team (STScI/AURA).
Acknowledgment: M. Shara and D. Zurek (AMNH), and F. Ferraro (ESO).

but in the scientific sense of a core scientific theory, one of the big explanations that form part of modern science. The big bang theory has transformed
astronomy from a simple description of what the universe is into a science
that can provide some explanations of how the universe changes over time.
The big bang theory has been described in many popular books. One recent
theory that does justice to the evidence is Coles (2001). Another, which deals
with the very recent issue of whether the cosmic expansion is accelerating, is
Kirshner (2002).

A BRIEF HISTORY OF THE UNIVERSE

Fourteen billion years ago, it all began. At the ultimate zero time of cosmic
evolution, the theory suggests that the temperature was infinite, another way

of saying that we cannot use our understanding of physics really that far. But after one-trillionth of a second (10^{-12} second), the universe was still extremely hot, but cool enough so that we can confidently say we understand the laws of physics that made it work. At this time, everything that we now see in our observable universe was compressed into a tiny ball about the size of the period at the end of this sentence. This ball, and the space it occupies, expanded very rapidly, and cooled as it expanded. As is found in any hot substance, light is present along with the matter, which is expanding.

During the first twenty minutes, nuclear reactions took place in the early universe. Protons and neutrons smashed together, forming helium nuclei when two of each were glued together. Some small quantities of deuterium, or heavy hydrogen, were left over from this process. The usual hydrogen atom has a bare proton as a nucleus. Deuterium has one proton plus one neutron. Small quantities of other odd nuclei, like Helium-3 and lithium, were also made at this time. And the universe continued to expand and cool.

At the million-year mark, the electrons in the universe combined with the atomic nuclei that were around, and the entire universe became composed of neutral particles. Light, instead of interacting with the charged atomic nuclei and the oppositely charged electrons, stopped interacting with anything. It just traveled through the expanding universe. As the universe expanded, its wavelength became longer and longer. By now, this light is in the microwave part of the spectrum. It is the same kind of radiation that, when more intense and more focused, can boil water so you can make a pot of tea.

At some time in the next billion years, structure began to form in the universe. The early universe contained the familiar elements hydrogen and helium. According to the current consensus model (see Kirshner 2002 for references), the universe also contained unfamiliar substances that astronomers call *dark matter* and *dark energy*. We have no idea of what kinds of particles make up the dark matter and dark energy. Our current understanding of how galaxies form has resulted in a consensus among astronomers that these strange entities exist. The structure in the universe consists of huge galaxies, some of which resemble the beautiful spiral galaxy shown in Figure 57.2.

THE BIG BANG AND STEADY STATE HYPOTHESES

The big bang theory was originally developed by Abbé Georges Lemaitre in the 1930s as the most logical explanation of the expansion of the universe. A series of measurements by Vesto Slipher of the Lowell Observatory and the better known Edwin Hubble of the Mt. Wilson Observatory had demonstrated in the 1920s that all galaxies beyond the five nearest were all moving away from the Earth. Follow this expansion backward in time, and you come to a dense beginning of the universe.

A group of astronomers in Britain developed an alternative explanation for the expanding universe in the late 1940s. This picture visualizes that matter is continuously being created as the universe expands. A force field associated

Spiral Galaxy NGC 3370

Hubble Heritage

FIGURE 57.2   A galaxy, or collection of billions of stars. If we could see our home galaxy from a distance, it would probably look somewhat similar to this galaxy.

Image credit: NASA, the Hubble Heritage Team and Adam Riess, and Space Telescope Science Institute.

with the creation of this matter pushes the universe apart, and the newly created matter then forms galaxies like those shown in Figure 57.2. There is no hot beginning. Its originators followed the very common practice, at least in the physical sciences, of naming this proposal the *steady state theory*, using the word *theory* in its popular sense and unfortunately adding to the popular confusion about this most important concept (steady state is actually a hypothesis, or scientifically untested assumption).

COSMIC FOSSILS: EVIDENCE THAT THE BIG BANG THEORY IS CORRECT

An all too common tendency among textbook writers is to present a story and simply leave it at that. What makes the big bang theory a core scientific theory rather than just a good idea is that there is a lot of evidence to support it. Even a brief article like this should refer to the evidence.

A prediction of all big bang cosmological models is that about one-quarter of the universe should be transformed from hydrogen into helium. About thirty years ago, a number of astronomers, including the author of this article, began to measure the abundance of helium in stars (for one of the older articles, see Shipman and Strom 1970). Since then, both the predictions and the observations have been refined, so that "about one-quarter" has become "23% with an uncertainty of 1%" (Pagel 1997). Our confidence that the agreement between model and observation is meaningful has grown considerably.

The light in the early universe, now transformed to microwave radiation, has been observed, measured, remeasured, and analyzed in great depth. Its accidental discovery by Arno Penzias and Robert Wilson, who were doing radio astronomy for Bell Telephone Laboratories in the 1960s, eventually won them a Nobel Prize. Their early work led to a blizzard of observational activity that included two NASA spacecraft and plans for a third. The big bang model makes very precise predictions of the distribution of this energy with microwave frequency. Observations confirm those predictions to incredibly high precision, about 10 parts per million. The detailed confirmation of the big bang model is generally regarded by astronomers as the most convincing piece of evidence in favor of the big bang theory.

An area of current research is the question of where the structure in the universe comes from. We live in a very rich universe that contains clusters of galaxies, stars, planets, and people. A complete theory of cosmology should not only explain where the universe came from but also account for the development of structure within it. A million years after the big bang, the universe was a smooth sea of gas. A thousand million years later, structure had begun to form in the universe, and pictures of the very early universe show many, many faint galaxies (see Figure 57.3).

TEACHING THE BIG BANG THEORY

With some attention to pedagogical matters, an instructor can lead students through our cosmological story and develop student understanding. A simple recital of the big bang theory is easy to do, but instruction and curriculum that are confined to storytelling lead to students who can simply recall the facts, performing at level 1 of Bloom's famous taxonomy of educational objectives (Bloom 1956). There are some aspects of the big bang theory that make it a relatively good vehicle for teaching the nature of science. My science education colleagues and I investigated student understanding in an astronomy course I taught some years ago, and some recommendations emerging from those results are described here.

EVIDENCE AND INFERENCE

Perhaps the most important aspect of the nature of science that we can teach is the relationship between evidence and inference (Osborne et al.

FIGURE 57.3   **The tiny blobs in this picture show galaxies, which formed at the beginning of time. This "Hubble Ultra Deep Field" shows a part of the sky where ordinary telescopes on the ground see absolutely nothing. The power of the third-generation cameras on the Hubble Space Telescope reveals nearly 1,000 galaxies, which formed very early in cosmic evolution.**

Image credit: National Aeronautics and Space Administration and European Space Agency; data obtained by R. Windhorst (Arizona State University), S. Yan (Spitzer Science Center, Caltech), and S. Beckwith (Space Telescope Science Institute). For more information about this image, see http://hubblesite.org/newscenter/newsdesk/ archive/releases/2004/07/text/.

2003). This concept is a key part of critical thinking (Paul and Elder 2001) and is often mentioned among key general education skills. However, not all scientific theories are equally amenable to teaching this most important understanding, which is a key part of critical thinking generally as well as being important to the scientific habits of mind. Our studies of my college astronomy course (Brickhouse et al. 2001, 2002) demonstrated that students had a relatively easy time understanding how the discovery of the microwave background radiation demonstrated that the universe was once hot and dense. The connection between evidence and inference in other areas of astronomy, such as the life cycles of stars, is much more indirect.

Students need to do more than just hear that there is a big bang theory and that discoveries of the cosmological helium and the microwave background provide evidence in support of it. One strategy I have used is to have students play the role of advisors to NASA who are asked to choose which of two or more possible missions should be supported by the agency. Is it more important to investigate the microwave background radiation or the helium abundance of very old objects? I have also asked them to play the role of prize judges who would award a substantial amount of money to those who discovered particular pieces of evidence. Yet another strategy is to put the students in the role of newspaper editors, who have to decide whether to run a particular newspaper story. For some results of the newspaper exercise in an astronomical context, see Shipman (2000).

NATURE OF SCIENTIFIC THEORIES

Testing competing scientific theories provides an excellent way of discussing what theories are. A nice feature of using the steady state hypothesis as a foil for the big bang theory is that it is possible to describe how the steady state works in about five minutes of class time. Students can understand the steady state hypothesis without a great deal of intellectual effort. There have been other areas of science in which incorrect historical theories are brought up, such as chemistry and biology. However, students have to work hard to figure out how phlogiston works in chemistry or how Lamarckian evolution differs from Darwinian evolution. It has always seemed to me that asking students to put considerable effort into understanding wrong ideas is at best misguided pedagogy. In the case of Lamarckian evolution, it may even lead to incorrect student conceptions.

However, simply presenting the nature of theories in a rather abstract lecture, and dealing with the steady state hypothesis and big bang theory four weeks later, led to some improvement of student understanding of what a theory is, but not the understanding that I as an instructor would have liked (Dagher 2004). Since we did the research that has led to this paper, I have changed my instructional strategies somewhat. I now include at least one more theory in the course structure. I am more explicit about defining a core scientific theory. I also make the connections between this definition and the theories that we treat more explicit. I have introduced the term *core scientific theory* to clarify the distinction between major scientific explanatory ideas and hypotheses that may or may not be right. So far, this term has not caught on outside of my own pedagogy, but I still find it to be useful.

SCIENCE AND RELIGION

About ten years ago, I started including a very brief treatment of science and religion in my science courses. It varies somewhat over the years, but it generally involves having the students write a short paper at home and then participate in a class lecture/discussion. The thrust of both exercises is that the big bang cosmology is most decidedly compatible with a theistic notion of God as is found in the world's three great monotheistic religions of Christianity, Judaism, and Islam. This surprisingly brief pedagogical intervention has had remarkable positive effects and absolutely no negative effects (Shipman et al. 2002).

SUMMARY

The big bang cosmology, in addition to attracting considerable student and popular interest in its own right, is an excellent way for teaching the scientific habits of mind. The relatively close connection between evidence and inference makes it easy to teach the importance of evidence, of data, of experiment, and of observation. Depending on what the instructor wants to do, a classroom treatment of the big bang can open up some rather deep issues.

ACKNOWLEDGMENTS

My astronomical work has been supported by various NASA Guest Investigator programs for the Hubble Space Telescope and the Far Ultraviolet Spectroscopic Explorer, and my scholarly work integrating science research and science education research is supported by the National Science Foundation's Distinguished Teaching Scholars Program (DUE-0306557).

REFERENCES

Bloom, Benjamin S. 1956. *Taxonomy of Educational Objectives: The Classification of Educational Goals.* New York: Longmans, Green.

Brickhouse, Nancy W., Zoubeida R. Dagher, Harry L. Shipman, and Will J. Letts IV. 2001. Diversity of Student Views about Evidence, Theory, and the Interface between Science and Religion in a College Astronomy Course. *Journal of Research in Science Teaching* 37:340–62.

———. 2002. Evidence and Warrants for Belief in a College Astronomy Course. *Science & Education* 11:573–88.

Coles, Peter. 2001. *Cosmology: A Very Brief Introduction.* Oxford: Oxford University Press.

Dagher, Zoubeida R. 2004. How Some College Students Represent Their Understanding of Scientific Theories. *International Journal of Science Education* 26:735–55.

Kirshner, Robert P. 2002. *The Extravagant Universe: Exploding Stars, Dark Energy, and the Accelerating Cosmos.* Princeton, NJ: Princeton University Press.

Osborne, Jonathon F., Mary Ratcliffe, Sue Collins, Robin Millar, and Rick Duschl. 2003. What "Ideas-about-Science" Should Be Taught in School Science? A Delphi Study of the "Expert" Community. *Journal of Research in Science Teaching* 40:692–720.

Pagel, Bernard E. J. 1997. *Nucleosynthesis and the Chemical Evolution of Galaxies.* Cambridge: Cambridge University Press.

Paul, Richard W., and Linda Elder. 2001. *Critical Thinking*. Upper Saddle River, NJ: Prentice Hall.

Shipman, Harry L. 2000. Teaching Astronomy through the News Media. *Physics Teacher* 38:541–42.

Shipman, Harry L., Nancy W. Brickhouse, Zoubeida R. Dagher, and Will J. Letts IV. 2002. Changes in Student Views of Religion and Science in a College Astronomy Course. *Science Education* 86:526–47.

Shipman, Harry L., and Stephen E. Strom. 1970. Helium Lines in O- and B-Type Stars. *Astrophysical Journal* 159:183–92.

58

Hurricanes and Tornadoes

Paul H. Ruscher

Many a child and adult are fascinated when nature displays its most awesome powers in the form of windstorms, blizzards, floods, high waves, volcanic eruptions, earthquakes, and tsunamis. The teachable moment is not always at hand, however, which would allow us to develop critical inquiry skills in real time under the threat of an impending natural disaster. Nevertheless, there are many scenarios that allow for the use of natural disaster material to forward educational goals and allow students to develop critical thinking and analysis skills at the same time.

When teaching meteorology, the tendency among many educators is to emphasize the *dramatic* natural disaster material, as this might be the only material that seems to motivate the students. Before moving on to discuss new ideas related to tornadoes and hurricanes, let's dispel that notion by stating that weather (and climate) can be made interesting without having to resort to discussions of "life and death struggles" and other sorts of natural calamity. Nevertheless, you will find the typical meteorologist more than casually interested in the outcome of a landfalling major hurricane or the development of an F5 tornado! So let's examine hurricanes and tornadoes!

THE CYCLONE

Any information presented to a student on hurricanes or tornadoes will lack the detail necessary unless material is also presented on *low pressure*, and what that means. In particular, our storms of interest are associated with low-pressure systems (*cyclones*, in the most generic use of the word). This generally could follow material on the composition of the atmosphere and discussions centered on the nature of air as a mixture of gases that has mass. Discussions like this lead to concepts such as density and pressure, the latter of particular importance in meteorology, following the experimental work of Galileo Galilee and Evangelista Torricelli in the 1600s, and the later observations of the movement of storm systems by Benjamin Franklin in the 1700s. Some excellent materials are available for teachers at the University of Illinois' WW2010 online meteorology textbook website (University of Illinois n.d.). Some time spent on background material on the nature of air, atmospheric pressure, and cyclones will reward anyone who is trying to delve into shared experiences on natural disasters such as hurricanes and tornadoes.

One of the important fundamental issues in science education is the notion of *scale*—how large or small is something? How long can something last? How do I know if my answer is in the right ballpark? Cyclones exist on many scales in the real atmosphere (Orlanski 1975), but most commonly we refer to the winter low-pressure systems so common in the middle latitudes as occurring on the synoptic or *macroscale*. These cyclones last days to weeks and occur over thousands of kilometers. On the *mesoscale*, we find storm systems that may last only a few hours to days or longer, but are much smaller in scale, with storm sizes on the order of a few hundred kilometers (such as the hurricane), down to the thunderstorms that are only a few kilometers to tens of kilometers across. The tornado is an indication of the atmosphere's tiniest of cyclones, with typical diameters of a few hundred meters or much less, at the *microscale*. Each of these phenomena in turn also has characteristic pressure ranges and wind speed ranges that must be understood in order to realize the complexity of these systems. Providing some practical information related to appreciation for scale is important in geoscience lessons, so that students know the expected outcomes of experiments or critical inquiry processes.

DISASTER MITIGATION

Whenever discussions on life-threatening situations are presented, it is advisable to present information on disaster mitigation as well. The 2004 Atlantic hurricane season was full of real-life examples of dangers felt by society, with a record-breaking five landfalling tropical cyclones in Florida alone (Tropical Storm Bonnie, Major Hurricane Charlie, Hurricane Frances, Major Hurricane Ivan, and Hurricane Jeanne). The costs of these natural disasters are typically reported in numbers of people killed or injured, and/or in millions (or billions) of dollars of damage. A better grasp of these tragedies requires an understanding of the science associated with natural forces that can cause such devastation. Case studies can be developed on individual storms in which students are brought through a series of experiences or intellectual activities that allow them to make critical decisions. An example is

"Stormy Weather" by Stephanie Stevenson (n.d.), which allows students to experience the "Perfect Storm" of 1991 (Junger 1997; Greenlaw 1999) through a Web Quest, among other activities. Students may be more amenable to examining natural disasters if, in addition to the playing of videos of the devastating and awesome power of nature, critical examination of the real effects on people and places is discussed and shown.

Some discussion on actions that should be taken under these threats is important, and, fortunately, there are some good materials that have already been developed. Project Safeside, a collaboration between the Weather Channel and the American Red Cross, teaches how individuals and communities can implement proper disaster preparedness in their own communities. A role-playing interactive simulation entitled Hurricane Strike! was developed for the Federal Emergency Management Administration (FEMA) by the University Corporation for Atmospheric Research (UCAR) COMET program and won the prestigious Battan Book Award of the American Meteorological Society (AMS) for 2003. In this simulation, a family is faced with decisions regarding how to get everyone to safety as Hurricane Erin approaches their home on the Florida coast. The American Red Cross also has an excellent new series entitled Masters of Disaster that teaches the concept of mitigation appropriately for various natural disasters, including hurricanes. Depending on the level of the class, questions and discussions can be raised about the desirability of living on the coast versus the reality of repairing and reconstructing damaged or destroyed homes. Who should pay for these damages, the government or the property owner? To what extent does individual responsibility come into conflict with government policy and/or public safety? Recent examples of mudslides in California and the devastating tsunamis in the Indian Ocean also come to mind here as great examples of natural disasters that can become teachable moments, later, in a case study approach. (To learn more about the resources discussed in this paragraph, see the website list at the end of this chapter.)

HURRICANES

What are hurricanes? Hurricanes are merely the strongest type of tropical low-pressure system found on Earth, at least in the Atlantic Ocean and East Pacific Ocean. Elsewhere they are sometimes called other names—and generically, meteorologists use the term *tropical cyclone*. Classification is via the Saffir-Simpson scale that is designed to estimate damage potential; it was created by a meteorologist and a building engineer. A typical hurricane life cycle begins with a swirl of clouds over the tropical ocean that becomes organized into a low-pressure system with embedded thunderstorms and heavy rain showers. At this stage, the feature may be classified as a *tropical wave* or a *tropical depression*. The Tropical Prediction Center in Miami, Florida, a branch of the National Weather Service of the U.S. National Oceanic and Atmospheric

Administration (NOAA), is responsible for tracking these features during hurricane season (June 1 through November 30 each year in the Atlantic, although storms often form outside of this window). When low pressure intensifies (drops) enough so that sustained winds reach 39 mph (34 knots) or greater, the storm is named as a *tropical storm*, according to the list of names in use each year. A *hurricane* is formed when sustained winds reach 74 mph (64 knots) or greater. Major hurricanes (category 3 or higher on the Saffir-Simpson scale) have sustained winds of greater than 115 mph (100 knots).

Typically, interactions with land, cold ocean water (temperature below 80°F or 26.5°C), dry air aloft, or wind shear (vertical changes in wind speed or direction) will weaken storms, and the absence of them provides a favorable environment for development or intensification. There are several ways in which the atmosphere behaves that make hurricanes and tornadoes likely systems for teaching by contrasting their fundamental processes. The most important aspects as far as their potential to cause problems are related to their forecast intensity, measured in terms of central pressure or highest wind speed; their forecast track (the path they will move); and their forecast size. All of these elements form the basis of the advisories that are issued by the Tropical Prediction Center every six hours when a storm is active. More frequent advisories are issued as needed for storms nearing land areas.

Hurricanes affect very large areas with strong winds that may last for hours, intense rainfall that may produce inland flooding, and coastal storm surges and storm tides that may wreak havoc on coastal structures, barrier islands, and beaches in general. These devastating effects may be much more intense in the interior of the storm, nearest the "eye wall" of the hurricane. This is one way in which hurricanes can be contrasted with tornadoes. The scale of these storms in terms of their size is quite large—both in terms of the area covered and the time over which storm conditions are experienced. Tornadoes typically move very quickly on the ground, and they are very small-scale phenomena. Thus the typical tornado affects a given area for only a few minutes at most, yet the effects can be just as devastating as a hurricane's, if *your* community or neighborhood is hit.

Another contrast based in science relates to wind shear. The hurricane is a large atmospheric heat engine that requires no interruption in its attempt to take warm moist air near the surface and lift it to the top of the troposphere, continuously releasing *latent heat* in the process. Any wind shear present in the environment can interrupt this process, thus limiting growth and intensification of the core of the storm. Recently scientists have begun to classify storm intensification processes by quantifying the amount of wind shear present in the storm environment and found very satisfactory results. These developments would not be possible without new computer software programs successfully being developed based on weather satellite imagery and use of other satellites such as satellite wind sensors and satellite-based radar. These allow for the continuous or near continuous monitoring of the ocean areas where these storms form, in the absence of any direct measurements by weather stations or human observers, since these developments usually occur over the open ocean where only an occasional ship may stray.

In recent years, new technologies based in research and operational reconnaissance aircraft that penetrate hurricanes have revealed some interesting new ideas about storm structure and the role that the storm environment plays in shaping changes in storm track and intensity. Many of these technologies are deployed by scientific staff working at NOAA's Hurricane Research Division in Miami, Florida, work done in conjunction with scientists from other government labs, university programs, and private industry. These new areas of research promise to provide better analysis of the inner structure of the storm and its effects on the ocean, which promises to reveal better forecasts of these storms as time goes on. New numerical models being built on supercomputers now have the capability of forecasting hurricane track five days in advance, which is superior to the three-day forecasts of only a decade or so ago. One such example is the hurricane superensemble forecasts issued by Florida State University for the 2004 Atlantic hurricane season, which provided the best overall guidance of any computer model.

These storms present intriguing possibilities for schools, since so many of them are used as evacuation centers, or shelters of last resort. In recent years, a false forecast landfall of Hurricane Floyd in 1999 led to the largest peacetime evacuation in the United States up to that time (the sequential landfalling storms in Florida in 2004 combined to exceed that number). Such questions as "Why do people decide to leave?" or "Why do people decide to stay?" come to mind, as well as "What leads emergency planning officials to decide to issue recommended or mandatory evacuation orders?" Such questions can take science into the more natural realm of inquiry for people, by focusing not on the difficult science questions at hand but on the response of people to such conflicts.

TORNADOES

In a typical year, approximately 1,000 tornadoes will affect the United States, the nation with the largest number of tornadoes in the world thanks to a unique combination of geography, oceanography, and weather/climate patterns. In recent years, much has been learned about the tornado, as scientists have begun to produce documentation at better and better scales (higher resolution). Beginning in the 1970s, newer Doppler radar technologies began to be deployed that revealed not only the intense precipitation with thunderstorms but also the winds inside a thunderstorm; and for the first time, radars some tens of kilometers away from a thunderstorm could visualize the rotating updraft area of a thunderstorm that so often serves as a harbinger of a developing tornado. Numerical simulations of these rotating updrafts have also been recent successes in the computer laboratories, where scientists have designed thunderstorm-scale computer renditions of the atmosphere, designed to simulate the change of a beautiful, fair weather, cumulus-filled sky into a raging thunderstorm producing hail, violent winds, and tornadoes.

Tornadoes are violently rotating columns of air associated with thunderstorms, which have a high degree of wind shear and instability associated with them. The instability is thermodynamic in nature, and occurs when warm moist air is found beneath much colder, denser air above. The wind shear is present normally because the warm moist air at low levels originates over the Gulf of Mexico; southerly winds generally bring it onshore and move it northward, where the air may come into contact with fronts or other features that force it to rise, meeting airstreams from other regions such as continental dry air from northern Mexico or cold, dry air from the northern Great Plains. Tornadoes are most common across the nation in the spring, but may form in any season. The most interesting events for meteorologists as well as the general public are the *tornado outbreaks*, which occur much like outbreaks of a virus. Conditions are ripe for the systems to develop, and symptoms may vary from no tornado in some storms to areas in which thunderstorms do not even form. However, in some instances, the ingredients are there for a particularly virulent strain, and an F4 or F5 tornado may form. Such a tornado hit Moore, Oklahoma, a suburb of Oklahoma City, in May 1999. Among the forty-five fatalities, it has been documented that many were killed because they did what they probably should not have done, that is, outrun the tornado. Owing to the strong wind shear and strong winds aloft on most typical tornado days, tornadoes usually move very fast along the ground. As such, even on the straightest roads with excellent visibility, they may be difficult to outrun. And yet, with all of those fatalities, not a single school-age child was killed (Brooks and Doswell 2002). In fact, it was speculated that thanks to the outreach efforts by public safety and weather officials in that area, schoolchildren often were responsible for enacting proper emergency procedures to stay safe (a tornado warning was in effect at the time). This is an outstanding example of how what is taught and experienced in school can have real-life consequences on life-saving activities.

FORECASTING EXTREME EVENTS

Much of the latest and greatest research done on both hurricanes and tornadoes is easily discovered on the Internet. The Tropical Prediction Center's website, www.tpc.ncep.noaa.gov, includes all of their current advisories, watches, and warnings (if any), as well as links to storm surveys and climatological reports. Experimental products, including five-day forecasts of wind radii begun in 2005, are also available. The Hurricane Research Division provides access to information on the hurricane reconnaissance aircraft program and their other research; find more information at www.aoml.noaa.gov/hrd/. The Storm Prediction Center (SPC) of NOAA is located in Norman, Oklahoma; SPC is responsible for monitoring the nation's weather for hazardous conditions such as severe thunderstorms, tornadoes, and winter storms. Climatological and research data as well as operational forecasts are easily retrieved from their website, www.spc.noaa.gov. (These and other helpful websites are listed at the end of the chapter.) Private sector meteorologists, usually employed by television networks, private companies, and individual television stations, also provide general weather forecasts and disseminate official weather information issued by NOAA to the general public. The most important forecasts disseminated are typically the watches and warnings, issued only by NOAA offices. A *watch* indicates that a particular threatening condition is possible, while a *warning* indicates that it is likely already occurring or is likely to occur. In recent years, there has been a push to allow private or commercial meteorologists to have more autonomy in their forecasts of extreme weather, while it is federal law that only NOAA entities may issue direct watches and warnings of these events. What consequences can you envision if the process of issuing watches and warnings were to be opened up to the private sector?

As this volume goes to press, the people of the Gulf Coast of the United States still reel from the losses associated with Hurricane Katrina, the most damaging storm ever to ravage the U.S. coastline. With losses estimated at over $75 billion, and over 1,600 confirmed dead and as many or more still listed as missing, Katrina is notable from many perspectives. Natural disasters of all sorts remain highly unpredictable. The details of Katrina's size, track, and intensity forecasts were filled with uncertainty three, four, and five days out, providing difficult decision points for people along the Gulf Coast; nevertheless the National Weather Service is widely praised as the only Federal agency that did its job well in the Katrina situation. Communities were in different states of readiness for this disaster in late August 2005, and as a result of an incapable levee system, many drowned not due to the direct effects of the storm, but due to failed levees from storm surge and/or wave action on inland Lake Pontchartrain adjacent to New Orleans as weakening Hurricane Katrina made landfall along the Louisiana/Mississippi state line. The period of 2004 and 2005 have brought two years of devastation to the southeastern United States, and many residents are beginning to realize with great clarity the threats that these storms can pose, yet property values continue to spiral upwards, and coastal populations swell, putting more people and places at risk. Our natural environment is suffering, as well, with coastal shorelines and wetlands that form natural barriers protecting inland areas disappearing. Notable hurricanes have struck the U.S. mainland from Texas to Maine, and it is important for any coastal residents and those dozens of miles inland to realize the power and devastation that these storms can cause. Teachers can provide a focal point for readiness and awareness by incorporating natural disaster materials appropriate for their area into their teaching. Statistics and other data provide useful ways to teach mathematical concepts including using data to make graphs and charts. Students can be encouraged to write about their feelings and carry out research on topics related to storms and disasters. Geographic awareness is developed by students' mapping out disaster areas and learning how to read flood zone maps and disaster declaration areas. A disaster like Katrina can not be avoided completely, but with a better educated populace, its effects can be significantly lessened.

REFERENCES

Brooks, Harold E., and Charles A. Doswell III. 2002. Deaths in the 3 May 1999 Oklahoma City Tornado from a Historical Perspective. *Weather and Forecasting* 17:354–61.

Greenlaw, Linda. 1999. *The Hungry Ocean: A Swordboat Captain's Journey.* New York: Hyperion.

Junger, Sebastian. 1997. *The Perfect Storm: A True Story of Men against the Sea.* New York: W. W. Norton.

Orlanski, Isidoro. 1975. A Rational Subdivision of Scales for Atmospheric Processes. *Bulletin of the American Meteorological Society* 56:527–30.

Stevenson, Stephanie. N.d. Stormy Weather. Education Central. www.educationcentral .org/stormy/body.htm.

University of Illinois. N.d. WW2010: The Weather World 2010 Project. University of Illinois, Department of Atmospheric Sciences, Urbana-Champaign. http:// ww2010.atmos.uiuc.edu/(Gh)/home.rxml.

WEBSITES

Bad Meteorology: www.ems.psu.edu/~fraser/BadMeteorology.html
EarthStorm at Oklahoma: www.mesonet.org/projects/earthstorm.php
EXPLORES! at Florida State University: www.met.fsu.edu/explores/
Florida EOC: www.floridadisaster.org
GLOBE program: www.globe.gov
Hurricane Research Division (HRD): www.aoml.noaa.gov/hrd/
Hurricane Strike! http://meted.ucar.edu/hurrican/strike/
Masters of Disaster: www.redcross.org/disaster/masters/
Project Atmosphere: www.ametsoc.org/amsedu/Proj_ATM/
Project Safeside: www.weather.com/safeside/
Storm Prediction Center: www.spc.noaa.gov
Stormy Weather: www.educationcentral.org/stormy/body.htm
Tropical Prediction Center (TPC): www.tpc.ncep.noaa.gov
WW2010: http://ww2010.atmos.uiuc.edu/

59

Global Warming

Miyoun Lim and Loaiza Ortiz

Global warming refers to an increase in the average temperature of the Earth's surface over time, which consequently causes global climate change. The gradual increase of the temperature of the Earth's lower atmosphere is attributed to the increased anthropogenic (human-caused) emission of greenhouse gases into the atmosphere.

GREENHOUSE EFFECT

The greenhouse effect is a natural process that warms the Earth's atmosphere. It occurs due to the presence of gas molecules in the atmosphere that absorb and trap heat that is radiated by the Earth's surface. These gas molecules, including water vapor (H_2O), carbon dioxide (CO_2), methane (CH_4), and nitrous oxide (N_2O), are called *greenhouse gases*.

Energy from the sun travels to the Earth primarily as a form of visible light. About half of the energy that passes through the Earth's atmosphere reaches the surface of the Earth and is absorbed. When the surface of the Earth eventually reemits the solar energy into the atmosphere, it is converted into heat radiation. Greenhouse gases present in the atmosphere trap and retain some of this heat radiation in the atmosphere. The greenhouse gases allow the incoming (short-wavelength) sunlight to pass through. However, once the energy is converted into heat (long-wavelength) radiation, the greenhouse gases keep the heat energy trapped in the atmosphere. Most of this re-radiated energy is reflected back to the Earth's surface. This process further warms the atmosphere of the Earth. In this process, the greenhouse gases act like the glass panels of a greenhouse, trapping re-radiated energy and keeping the heat in. This is where the name *greenhouse effect* comes from. Due to this natural process, the Earth's average surface temperature is about 16°C, which makes life on Earth possible. Without the greenhouse effect, the average global surface temperature would be much lower and life would be unsustainable.

ENHANCED GREENHOUSE EFFECT

The natural greenhouse effect provides a hospitable environment for life on Earth. However, the greenhouse effect raises concern when human activities contribute to the significant increase of greenhouse gases in the atmosphere and consequent global climate change. Given greenhouse gases' long atmospheric lifetimes (decades to centuries), increased anthropogenic emissions of greenhouse gases result in their accumulation in the atmosphere.

According to the Intergovernmental Panel on Climate Change (IPCC 2001), the average surface temperature of the Earth has increased about 0.6°C over the last 100 years. Since the Industrial Revolution, and particularly over the last fifty years, human activity has altered the composition of the atmosphere and contributed to increased concentrations of greenhouse gases therein. Scientists agree that increased inputs of greenhouse gases into the atmosphere from human activities enhance the Earth's natural greenhouse effect and led to an increase of the average surface temperature of the Earth. The increase in temperature due to this enhanced greenhouse effect is known as *global warming*.

EVIDENCE

Considerable evidence shows that we are experiencing global warming. As mentioned above, the average temperature of the Earth's surface has risen about 0.6°C over the twentieth century, and most of the increase has taken place in recent decades. Direct atmospheric temperatures have been measured since 1861. Since then, it has been recorded that the 1990s was the warmest decade, and 1998 was the warmest year, followed by 2002 and 2001.

An examination of historic data also reveals evidence of global warming. Scientists have examined atmospheric carbon dioxide concentrations during different time periods by analyzing the gas bubbles trapped in the Earth's ice core. By examining these data over time, scientists found that levels of CO_2 concentrations and mean global temperature have been increasing (see Figure 59.1) since the end of the last ice age (about 10,000 years ago). The data also show that increased concentrations of CO_2 in the atmosphere are related to the increased ability of the atmosphere to trap heat, leading to the increase of the temperature of the Earth's surface. Scientists also agree on the correlation

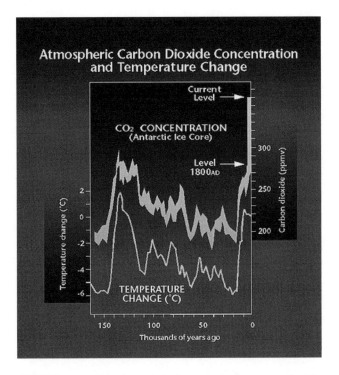

FIGURE 59.1 Historic record of atmospheric carbon dioxide concentration and temperature change from the Vostok ice core; see http://clinton5.nara.gov/media/gif/Figure5.gif.

between the increased use of fossil fuels, the increase of atmospheric concentration of greenhouse gases (e.g., carbon dioxide), and global temperature change.

ANTHROPOGENIC CAUSES OF GLOBAL WARMING

Many scientists, according to the IPCC report (2001), agree that "a discernable human influence" is responsible for this global warming, particularly due to the use of fossil fuels, industrial agricultural activities, and changes in land use. The use of fossil fuels is one of the primary sources of anthropogenic greenhouse gases. We use fossil fuels in transportation of people and goods, to heat homes and businesses, and for electricity. The use of fossil fuels releases huge amounts of greenhouse gases such as carbon dioxide (CO_2) and methane (CH_4) into the atmosphere. Carbon dioxide has increased 31 percent since 1750, from 280 parts per million by volume (ppmv) to more than 370 ppmv, half of the increase having occurred since 1965. This rate of increase is unprecedented. Coal-burning power plants are the primary source of CO_2 emission in the United States. Automobiles are the second largest source, creating nearly 1.5 billion tons of CO_2 annually.

POTENTIAL IMPACT OF GLOBAL WARMING

Global warming is a complicated phenomenon; it is hard to predict exactly how anthropogenic global warming will affect the Earth, how serious the effects will be, and how long they will last. However, each year scientists learn more and more about the effects of global warming on the Earth and develop better models to predict the potential outcomes. Even with the uncertainties, many scientists agree that without climate policies to help mitigate the trends, the average surface temperature of the Earth will increase by 1.4–5.8°C over two decades (1990–2010).

Scientists also agree on various potential consequences that are likely to happen if the current trends continue. These are presented below.

- Rise in the average temperature on Earth: the increase in the Earth's average temperature may cause climate pattern changes, including more frequent and intense heat waves and extremes of drying and heavy rainfall. These changes in precipitation patterns may increase the risk of drought, wildfires, and floods. The potential of extreme heat waves will also affect human and animal health.

- Sea level rise: the increase of the average global temperature may lead to the further rise of sea levels because water expands as its temperature rises. Furthermore, the warming will speed the melting of glaciers and ice caps. The potential rise of sea levels will cause coastal flooding, loss of coastal areas, and disruption of coastal ecosystems and human populations. Low-lying countries with large coastal populations will be in greater danger.

- Impact on human health: global warming may affect human health as well. Higher temperatures and extreme heat waves will increase the risk of strokes, heart attacks, and respiratory problems. Insect-borne diseases may spread to new areas.

- Impact on ecosystem: the global temperature rise will force plant and animal species to adapt and/or migrate. Plant and animal species that cannot adapt to climate change would face the danger of extinction. Global warming will disrupt ecosystems and will result in biodiversity loss.

Scientists predict that anthropogenic effect on climate change will last for centuries even after greenhouse gas concentrations are stabilized. Since greenhouse gases have a long lifetime in the atmosphere, anthropogenic increase of greenhouse gases will have a long-lasting effect on atmospheric composition and reaction. Also, sea level rise will continue for hundreds of years after stabilization of greenhouse gas concentration because the deep ocean needs a long timescale to adjust to climate change.

EDUCATION ABOUT GLOBAL WARMING

Educating about global warming is one of the most urgent and fundamental actions we must take to slow the effects of global warming. Education about global warming should include not only the scientific understanding of the phenomenon, but also understandings of its sociopolitical implications and ramifications. To nurture students' deep and balanced understandings of global warming, educators need to introduce and discuss uncertainties, differing opinions involved in scientific explanations, and predictions regarding global climate change and its consequences. Also, global warming could appear to be an intangible and disempowering topic for students due to its scale and the magnitude of its potential impact. It is important to discuss action strategies for students to engage in that could help slow the effects of global warming. By exploring personal, social, and political action strategies, students can develop a deep understanding of the phenomenon and feel empowered by participating in the processes of slowing the effects of global warming.

REFERENCE

Intergovernmental Panel on Climate Change (IPCC). 2001. *Climate Change 2001: The Scientific Basis: Contribution of Working Group I to the Third Assessment Report of the Intergovernmental Panel on Climate Change*. Cambridge: Cambridge University Press.

SUGGESTED READINGS

Cunningham, William P., and Mary Ann Cunningham. 2002. *Principles of Environmental Science: Inquiry and Applications*. New York: McGraw-Hill.

Enger, Eldon D., and Bradley F. Smith. 2002. *Environmental Science: A Study of Interrelationships*. Boston: McGraw-Hill.

Karl, Thomas F., and Kenneth E. Trenberth. 2003. Modern Global Climate Change. *Science* 302:1719–23.

Miller, George Tyler. 2004. *Environmental Science: Working with the Earth*. Pacific Grove, CA: Brooks/Cole.

60

Environmental Protection

Loaiza Ortiz and Miyoun Lim

In this chapter, the issue of environmental protection is approached from the perspective of sustainability. We begin with a discussion of the history and definition of sustainability—what does sustainability mean within the context of environmental protection? And how should it shape the ways in which we choose to live our lives? We follow with a discussion of key environmental protection issues framed around strategies for sustainable living, and conclude with suggestions for incorporating these issues into science teaching and learning.

In the late 1880s, John Audubon led the American conservation movement grounded in the belief that our natural environment should be kept separate and protected from people. Rooted in the Christian values of reverence for creation, the movement hoped to inspire piety through awe of nature's beauty and diversity. In the beginning of the twentieth century, the main reasons for creating protected areas were to increase the public's access to beautiful natural areas for recreation and tourism, to protect the wildlife habitat for hunting purposes, or to protect forests from logging interests. Protecting areas from land speculators, and ensuring that certain areas were preserved for the benefit of all, were the officially stated objectives for creating parks. More recently in our history, the purposes for protecting land have changed.

In 1987, the World Commission on Environment and Development (WCED) published *Our Common Future*. The report was influential in a number of ways. It stated that critical global environmental problems were primarily the result of the enormous poverty of the Global South (so-called

Third World, or developing, countries) and the nonsustainable patterns of consumption and production in the North. The report alerted the world to the urgency of making progress toward economic development that could be sustained without depleting natural resources or harming the environment. It proposed the concept of sustainable development, a strategy that united development and the environment.

The report highlighted three fundamental components of sustainable development: environmental protection, economic growth, and social equity. It defined *sustainable development* as development that "meets the needs of the present without compromising the ability of future generations to meet their own needs" (WCED 1987, 8). The WCED emphasized that the environment should be conserved, and our resource base enhanced, by gradually changing the ways in which we develop and use technologies.

How we define a protected area has evolved along with such changes in environmental thinking. As the need to conserve habitat for endangered or threatened species gained support through the 1970s and 1980s, we saw a significant change from the preservation of areas for recreational opportunities to the conservation of areas in order to preserve natural features and biodiversity for future generations. *Our Common Future* was a turning point in such thinking. Its species and ecosystems chapter, beginning with the statement, "Conservation of living natural resources—plants, animals, and microorganisms, and the non-living elements on which they depend—is crucial for development" (WCED 1987, 147), discusses species loss and extinction, habitat alterations, causes of extinction, the economic values at stake, and specific actions that governments can take to protect species and habitats. It did not, however, define the term *protected* or identify specific land management objectives for protected status. As a consequence, the term *protected areas* has come to be used to designate a variety of areas with differing objectives and management regimes. For example, some protected areas are closed to natural resource and industrial development, whereas others allow such uses if they are compatible with the specific objectives and management practices of the area.

Land and other resource protection is just one piece of conservation that must be seen within the larger context of sustainable development. It is important to look beyond the quantity of land protected to the quality of biodiversity, to how the land is managed, to what the land and resource rights of the local people are, and to what the costs and benefits are and how they are spread across society. A focus on endangered species or species diversity must be balanced with attention to components of biodiversity that are valued by the local people; what are the species they value for food and medicine? What are the alternatives to keeping humans and nature separate? How else can lands be managed in ways that promote conservation and are not lands fenced in and guarded from human activity? How can we link conservation with livelihoods? How can we address local people's needs in planning and management of protected lands?

Sustainable development, an integration of conservation and development, is a balance between the need to preserve environmental resources and the necessity of economic progress and development. The tension between resource protection, conservation, and development is often misunderstood as an issue of "people versus nature." The notion of sustainable development challenges the thinking that says that protecting biodiversity means keeping nature and people separate. In order to live sustainably, we must reduce our demands on nature's resources and understand the tensions between sustainable use of renewable resources (such as wood) and preservation of nonrenewable resources (such as fossil fuels).

In 1992, the United Nations convened the Conference for Environment and Development in Rio de Janeiro, Brazil. The meeting, commonly referred to as the Rio Earth Summit, led to greater awareness about, and understanding of, the idea of sustainable development. One important focus of that meeting was seeking ways of living, working, and being that enable all people of the world to lead healthy, fulfilling, and economically secure lives without destroying the environment and without endangering the future welfare of people and the planet.

Sustainability is now thought of as more than an ecological concept; it is an economic, political, and social concept as well. Sustainable development concerns the development of societies and economies by focusing on the conservation and preservation of our environments and natural resources— environmental, economic, and social progress and equity within nature's limits. Though the concept has gained in popularity, its definition remains elusive. One definition of sustainable development is maintaining a quality of life for *all* within the limits of nature. Thus, society meets its needs while preserving biodiversity and natural ecosystems in such a way that allows future generations to maintain a quality of life.

Let us look at sustainable food systems as an example. A food system describes the steps involved in getting food "from the farm to the plate." A sustainable food system allows, as the above definition suggests, all members of society—from the farmer to the farm worker to the slaughterhouse worker to the consumer—to maintain a quality of life within the limits of nature. Among the most important aspects of a sustainable food system are organic or minimal-treatment farming techniques, crop species diversity over monoculture, living wages and humane working conditions for farm workers, and localized distribution structures that promote community, provide access to healthful food for all, and stimulate local economies. As consumers we have choices about what kind of food system we want to support. By buying food at farmers' markets, we are supporting local farmers and reducing fossil fuel demands; and by choosing organically farmed produce and meats, we are supporting techniques that reduce soil and water pollution and contribute to the health of our society. Thus, sustainable food systems integrate conservation of resources with economic development in ways that promote ecological and social justice.

Education for sustainable living is fundamentally about building and nurturing sustainable communities that do not interfere with nature's ability to sustain life. As science educators, we can accomplish this by educating for ecological literacy and infusing understandings about the connectedness of all life into the life and physical sciences. Seeing the world as an interconnected whole and understanding the fundamentals of ecology—networks, nested systems, cycles, flows, development, and dynamic balance—mean understanding nature's life-sustaining patterns and processes and our place within nature. Scientifically literate citizens are necessarily ecologically literate. Understanding today's social and environmental problems, including global warming; air, water, and soil pollution; deforestation; loss of farmland; loss of biodiversity; and health, overpopulation, poverty, energy consumption, waste production, and transportation issues is not enough. We must also understand how these issues are intricately connected to each other and to advances in science, technology, and economic progress, and the implications of individual and societal decisions for local and global environmental and societal health. Most importantly, teachers can help empower students to take actions at the individual and community levels that lead to sustainable living.

REFERENCE

World Commission on Environment and Development (WCED). 1987. *Our Common Future*. Oxford: Oxford University Press.

ADDITIONAL RESOURCES

Center for Ecoliteracy. 2005. www.ecoliteracy.org.
Cloud Institute for Sustainability Education. N.d. www.sustainabilityed.org.
Cunningham, William P., and Mary Ann Cunningham. 2002. *Principles of Environmental Science: Inquiry and Applications*. New York: McGraw-Hill.
Enger, Eldon D., and Bradley F. Smith. 2002. *Environmental Science: A Study of Interrelationships*. Boston: McGraw-Hill.
Goldman, Harriet. 1999. *Sustainability Education: Teaching Sustainability in Every Classroom*. http://ceres.ca.gov/tcsf/seg/.

61

Protecting the Air We Breathe

Miyoun Lim and Loaiza Ortiz

Air pollution refers to the presence of unwanted substances in the air we breathe that can be harmful to our health and our environment. Air pollution may come from both natural sources (e.g., natural forest fires and volcanic activities) and anthropogenic, or human-caused, sources (e.g., motor vehicles and power plants). Anthropogenic pollutants are considered more serious because they tend to be localized in populated areas and easily build up to harmful levels of concentration. In this chapter, we discuss air pollution in terms of anthropogenic air pollution.

COMMON AIR POLLUTANTS

Scientists classify outdoor air pollutants into two groups: primary and secondary pollutants. Primary pollutants are emitted directly into the atmosphere in a potentially harmful form from a given source. Carbon monoxide (CO), volatile organic carbons, particulate matter, sulfur dioxide (SO_2), and oxides of nitrogen are examples of primary pollutants. Secondary pollutants are formed through chemical reactions when some of the primary pollutants react with one another or with the other components in the atmosphere. For example, photochemical oxidants such as ozone (O_3) are formed when oxides of nitrogen react with volatile organic carbons under the sunlight.

There are several air pollutants that are commonly found in the United States that can harm public health and the environment. Under the Clean Air Act, the Environmental Protection Agency (EPA) designated six "criteria" pollutants (see Table 61.1) as indicators of outdoor air quality: CO, nitrogen dioxide (NO_2), SO_2, O_3, particulate matter, and lead (Pb). The EPA has set National Ambient Air Quality Standards for these six criteria pollutants as maximum concentration levels above which harmful effects on human health and the environment may occur. The EPA monitors the levels of criteria pollutants in the air and emissions from various sources.

EFFECTS OF AIR POLLUTION

Air pollution can be harmful to our health and our environment. Effects to our health could be acute (such as burning eyes and nose; itchy, irritated throat; and trouble breathing) or chronic (such as cancer, birth defects, brain and nerve damage, and long-term injury to the lungs and breathing passages). How air pollution affects different people depends on the intensity and duration of exposure, and on individual behavior and susceptibility. In general, more susceptible populations, such as children, the elderly, and people with preexisting health problems, especially respiratory and/or cardiovascular problems, are at greater risk. Air pollution can also have negative effects on our environment by disrupting the ecosystem balance and diversity, and on physical structures by causing damage to buildings and monuments.

COMMON AIR POLLUTION ISSUES

In what follows, we discuss some important issues related to air pollution and how they affect air quality, our health, and our environment.

Smog

There are two types of smog: industrial smog and photochemical smog. The infamous London smog of 1952 is an example of industrial smog (gray air), which consists of sulfur dioxide, sulfuric acid, solid particles, and aerosols. Burning large amounts of coal and heavy oil without proper control and regulation causes industrial smog. While industrial smog is rarely a concern in most developed countries, it still poses a problem in rapidly developing countries.

Photochemical smog (brown air), the smog in Los Angeles, is a mixture of numerous primary and secondary pollutants, such as from motor vehicles, industrial plants, and household products. Ozone (O_3) is a well-known photochemical pollutant that is formed when nitrogen dioxide reacts with volatile organic carbons under sunlight. Nitrogen dioxide (NO_2) is responsible for creating the brownish haze over many cities.

"Acid Rain"

Acid rain is a general term that refers to acid deposition on the Earth coming from the atmosphere. Sulfur dioxide (SO_2) and oxides of nitrogen are

TABLE 61.1   Six "Criteria" Air Pollutants

Pollutant	Anthropogenic Sources	Effects
	CARBON MONOXIDE (CO)	
• Odorless, colorless gas	• Motor vehicle exhaustion • Incomplete burning of fossil fuels • Cigarette smoking	• Reduced oxygen delivery to the body's organs and tissues • Damage to cardiovascular, respiratory, and nervous systems • Dizziness, headaches, and fatigue at higher concentration
	NITROGEN OXIDES (NO$_X$)	
• Light-brown gas	• Burning of fossil fuels in motor vehicle, power plants, industrial plants, and fertilizer	• Smog and acid rain • Lung irritation and damage • Increases susceptibility to respiratory infections • Exacerbates respiratory diseases such as asthma and chronic bronchitis • Damages trees and aquatic systems • Visibility reduction • Property damage through acid deposition
	SULFUR DIOXIDE (SO$_2$)	
• Colorless gas • Odorless at low concentration but pungent at very high concentrations	• Coal burning in power plants • Industrial processes (metal ore smelting)	• Acid rain • Causes respiratory distress and diseases • Exacerbates respiratory diseases such as asthma • Visibility reduction • Property damage through acid deposition

(continued)

TABLE 61.1 *continued*

Pollutant	Anthropogenic Sources	Effects
OZONE (O$_3$)		
• Colorless gas • Major constituent of photochemical smog, giving acrid, biting odor to it	• Motor vehicles and industries	• Irritates and impairs respiratory system • Lowers resistance to colds and pneumonia • Exacerbates chronic diseases such as asthma, bronchitis, emphysema, and heart disease • Damages vegetation and animal tissues
PARTICULATE MATTER		
• Mixture of solid particles and liquid droplets • Varies in sizes (coarse and fine particles)	• Burning coal in power and industrial plants • Burning fossil fuels • Agriculture • Unpaved road • Construction	• Finer particles can lodge deep in our lungs and cause lung damage and respiratory problems • Aggravates asthma and chronic bronchitis • Increases respiratory symptoms • Haze and reduced visibility • Affects diversity and balance of ecosystems
LEAD (Pb)		
	• Leaded gasoline • Paint • Smelter (metal refineries) • Lead manufacturer • Storage batteries	• Accumulates in the human body • Damages brain and rest of nervous system (children at greater risk) • Mental retardation • Harms wildlife

the primary causes of acid rain. Once these pollutants are emitted into the atmosphere, they react with water and other oxidants to form acidic compounds. These compounds remain in the atmosphere for several days. During this time, they precipitate to the Earth's surface in wet (rain, snow, and fog) or dry (gas and particles) forms.

Acid rain has harmful effects on human and ecosystem health (e.g., acidification of water and soils and damage to fish, wildlife, and forests) and causes damage to physical structures. Acid particles also affect air visibility.

Ozone Depletion

Ozone (O_3) can have positive or negative effects depending on where in the atmosphere it is found. It has negative consequences for our health and the health of the environment when it is present in the air we breathe. This is called *ground-level ozone*. Ozone has beneficial effects when it is present in the upper atmosphere, where it absorbs incoming ultraviolet (UV) radiation and protects living organisms on the Earth from harmful ultraviolet rays.

Over the past several decades, the protective ozone layer has been damaged. This is often referred to as the *hole in the ozone*. The most commonly cited example is the "hole" or thinning of the ozone layer over the Antarctic region. The principal causes of ozone depletion are chlorine-based aerosols such as chlorofluorocarbons (CFCs) and other halon gases, which are emitted from air conditioners, refrigerators, insulating foam, and some industrial processes. Since ozone depletion allows more UV radiation to come through the atmosphere, it causes UV radiation–related effects on our health and the environment, such as increased risk of skin cancer and cataracts, and suppression of the immune system. Ultraviolet radiation is also harmful to plant and animal tissues, and can have adverse effects on agricultural crop yields.

Indoor Air Pollution

When we think of air pollution, we often think of outdoor air pollution. However, indoor air pollution also affects our health, often more seriously than outdoor air pollution. The EPA has found that indoor concentrations of toxic air pollutants are often higher that outdoor concentrations. People spend more time indoors (about 90 percent) than outdoors. Thus indoor air quality becomes critical, due to the increased levels of exposure.

There are various sources of indoor air pollution including combustion sources (e.g., gas stoves, fire places, and cigarettes), building materials (e.g., asbestos), furniture (e.g., pressed wood products), cleaning products, heating and cooling devices, and outdoor sources (e.g., radon, pesticides, and outdoor air pollutants). Among these numerous indoor air pollutants, smoking is one of the most significant sources of indoor air pollution. The U.S. surgeon general estimates that 430,000 people die each year in the United States from emphysema, heart attacks, strokes, lung cancer, all related to smoking.

Sick building syndrome refers to indoor air pollution causing dizziness, headaches, coughing, sneezing, nausea, burning eyes, chronic fatigue, and flu-like symptoms. New buildings are often more "sick" than older ones due to the reduced air exchange and the chemicals released from new materials.

PROTECTING THE AIR WE BREATHE

Air pollution affects everyone, and as a society we must work to protect the air we breathe, indoors and outdoors. Education plays a key role in understanding the issues and concerns related to air pollution and the actions necessary to curb its effects on human and environmental health. A lot of pollution comes from power plants, industrial sources, and motor vehicles, so it may seem that our individual actions may have little impact. Thus, it is important for students to understand that the choices they make as individuals make a difference in the air we breathe every day. They have the power to change their home, transportation, and consumer patterns to help improve the air. Lessons that focus on the scientific concepts that give students deeper understandings of the phenomenon of air pollution must be integrated with discussions about taking action: choosing alternatives to driving such as bicycling, walking, and public transportation; using environmentally responsible products for home and office; and supporting the development and demand for products with fewer pollutants are some ways in which everyone can reduce air pollution.

SUGGESTED READINGS

Cunningham, William P., and Mary Ann Cunningham. 2002. *Principles of Environmental Science: Inquiry and Applications*. New York: McGraw-Hill.

Enger, Eldon D., and Bradley F. Smith. 2002. *Environmental Science: A Study of Interrelationships*. Boston: McGraw-Hill.

Miller, George Tyler. 2004. *Environmental Science: Working with the Earth*. Pacific Grove, CA: Brooks/Cole.

62

Childhood Overweight and Science Education

Isobel R. Contento, Angela Calabrese Barton, Pamela A. Koch, and Marcia Dadds

In this chapter we describe why childhood overweight should be an issue that is addressed in science education. We also propose a model that science educators can use for incorporating issues of childhood overweight into inquiry-based school science at the middle school level.

CHILDHOOD OVERWEIGHT: AN EPIDEMIC

Our involvement in the fields of nutrition and science education has made us fervently aware of the issue of childhood overweight.[1] The research is compelling: recent data suggest that over 50 percent of the population in the United States is considered to be overweight or obese, with the number of overweight children doubling in the past two decades.

The issue of "childhood obesity" has reached national consciousness, even though the rate has been increasing steadily over the years. Childhood overweight is a concern because it is associated with increased cholesterol levels and risk of type II diabetes. It is also a strong predictor of obesity later in life, where it is associated with metabolic syndrome, characterized by abdominal obesity, high blood lipids, high blood pressure, and elevated blood sugar. Childhood overweight also carries with it social consequences since overweight children often experience rejection by peers and other forms of social discrimination, dissatisfaction with their bodies, and low self-esteem.

PREVENTION OF CHILDHOOD OVERWEIGHT

Prevention of weight gain is easier, less expensive, and more effective than treating obesity after it has developed. The Surgeon General's 2001 *Call to Action* (U.S. Department of Health and Human Services 2001) includes a specific call for overweight prevention activities focused on healthy eating and physical activity in *schools*. Yet few projects exist that offer models for how schools might respond to the issue of childhood overweight. Many questions remain: how do we teach children about the challenges associated with childhood overweight in ways that are empowering, rather than overbearing? At what grade levels would such an intervention be most effective? How should any intervention program balance between helping children *understand* the problem and helping them *act* on the problem? Where should the program be housed—in physical education or health class? In special assemblies? Or as part of the regular academic curriculum?

HOW SHOULD SCIENCE EDUCATION RESPOND?

We believe that any school intervention focused on childhood overweight must go beyond previous interventions that have only been directed at a specific diet-related behavior such as fat reduction or fruit and vegetable consumption to teach children about how the body works and how bodies become overweight. Children also can benefit from having skills to analyze our current food and exercise environment and understand the dramatic changes that have happened in a short time period. We also believe that childhood overweight should be addressed across the school curriculum. However, we also think that science class offers a special venue because it is in science class where students can investigate through inquiry-based approaches how the body works and how overweight contributes to the body's well-being.

BECOMING COMPETENT THROUGH UNDERSTANDING OUR BIOLOGY, THE ENVIRONMENT, AND PERSONAL BEHAVIORS

In our work,[2] we have proposed to go beyond teaching children which behaviors to enact and the strategies to enact them, to include how to think about food and their behaviors in ways that help them feel competent in navigating today's food and activity environment. We believe that any childhood overweight prevention program has to focus on learning about the biology, the environment, and the personal behaviors influencing overweight.

Biology

Humans seem to be born with a liking for sugar and salt, and probably also for fat, and a rejection of bitter and sour flavors. Yet, humans are also born with the capacity to learn food preferences. Children can learn to like foods if they have positive interactions with them. Since high-fat, high-sugar, and high-salt foods are easily available and often served in positive social contexts, such as special occasions, young children come to like these foods due to their innate likings for these foods and the positive social context. In terms of quantity of intake, very young children are sensitive to caloric density differences between foods and adjust their intake accordingly, through a process of self-regulation. However, there is evidence that satiety, or the sense of feeling full, also becomes conditioned as children grow older.

Environment

What, then, is the food and physical activity environment in which children today have to make their choices, and what are the impacts? There are many statistics that show us just how much the social and economic environment has shaped how we eat. For example, in the last twenty-five years, the number of snacks eaten away from home has doubled, and most of these snacks are low in nutrient density and the drinks high in calories and added sugars. Furthermore, the average child sees 10,000 food ads per year, and the food industry spends $10 billion each year to influence eating behavior of children. Finally, children are much less active: young people aged 2–18 spend on average over four hours per day watching television and videotapes, playing video games, or using a computer. Walking and bicycling by children aged 5–15 dropped 40 percent between 1977 and 1995, partly due to lack of sidewalks and more use of cars. Clearly children have to make choices in an environment in which food is ubiquitous, tasty, cheap, and heavily advertised and being sedentary is easy.

Personal Behaviors

Studies have shown that for children, as for adults, taste preferences dominate food choice decisions and that family and social forces are also important, confirming that innate desires for sugar, fat, and salt and the food environment interact to influence food choice behavior. While the impacts are numerous, we think it is important to highlight three behaviors here. First, there is the increase in sugar and fat intake, due to increases in soda and fast-food consumption. Second, there is an increase in how much children eat due mainly to increasing portion sizes at restaurants (e.g., supersized). Third, there is an increase in sedentary lifestyles.

Becoming Competent as Middle Schoolers

It is clear, therefore, that any program focused on childhood overweight should teach children about their biology and why they have innate preferences for certain kinds of foods, as well as the ability they have to learn to eat in particular ways and how that ability is influenced by social context. We also believe that as students learn about the complex systems of their biology, the environment, and their behaviors, they should also learn how to competently navigate that system and that the best time to do this is in middle school (ages 12–14), as students are beginning to make more and more food choices for themselves.

While young children mainly make their food choices based on what they like, by middle school, cognitive-motivational processes also become important influences on food choice. Youth become more able to link cause and effect and to perceive the consequences of their actions. Thus, they can make food choices in light of their perceptions of health and weight concerns as well as taste and convenience. In terms of opportunity, by adolescence children receive $6 billion per year from caretakers and spend half of it on snacks. So middle school youth have both the capacity and opportunity to practice personal control and mastery in the area of food choices and physical activity.

Behavior, however, is motivated not only by the anticipation that the behavior will bring about desired outcomes, *but also by a strong sense of being able to exert personal control over the environment.* Children do have choices and need to learn to take control to create for themselves personal food and activity environments that are conducive to health and to body-size regulation. Therefore, we believe it is important to provide youth the opportunity to achieve a sense of personal control and mastery through understanding the complex systems of biology, the environment, and their behaviors and by learning how to take action based on their understandings. We call this personal control, mastery over their own behavior, and ability to create *personal* food and activity environments *competence.* We believe it is essential to assist youth to be able, through analyzing, understanding, and taking action in the current food system and sedentary environment, to acquire the skills and motivation to become what we call *competent navigators and creators* of their personal food and activity world. Thus, youth will be able not only to make choices within the system but also to "beat" the system, so to speak; to experience self-determination; and to gain personal satisfaction in achieving personal control and creating personal environments that are healthful and reduce risk of overweight.

CONCLUSIONS: A SCIENCE EDUCATION APPROACH

In short, we believe that in order to prevent childhood overweight, a combined approach must be taken. Children must have opportunities to investigate and develop deep conceptual understandings of the complex relationships between their biology, the environment, and personal behaviors. They also need adequate scaffolded opportunities to develop the scientific skills they need to "beat the system," or to become competent eaters and exercisers.

Our own approach has been to combine these two features in a twenty-four-lesson, inquiry-based curriculum called Choice, Control and Change. In this curriculum, the content investigations provide opportunities for students to understand their biology, the environment, and personal behaviors; classroom activities provide opportunities for students to explore their experiences and feelings, and to reflect on them through self-assessment of eating and activity patterns and learning about consequences; and use of our inquiry-based approach provides students the opportunity and skills to learn making choices and taking control.

Whatever one's approach is, one thing is clear: children will continue to be bombarded with opportunities to increase their fat and sugar intake cheaply and easily while remaining sedentary. This is not acceptable. We must find ways to work with individuals both inside and outside of school to help children gain the understandings and skills they need to competently navigate their worlds to make better and healthier choices.

NOTES

1. In children and adolescents, the term *overweight* is used rather than *obese*. Childhood overweight is defined as a body mass index (BMI) of greater than the 95th percentile on age- and sex-specific growth charts.

2. Our research is funded through the National Institutes of Health [RR020412].

REFERENCE

U.S. Department of Health and Human Services. 2001. *Surgeon General's Call to Action to Prevent and Decrease Overweight and Obesity.* Washington, DC: Office of the Surgeon General.

63

The Truth about Trans Fats Will Break Your Heart

Danielle Dubno

The next time you reach for a bag of chips, cookies, or crackers, look at the ingredients listed on the package. If you see the words *partially hydrogenated* or *shortening*, you can be sure that the food you are about to consume contains a special type of fat called *trans fat*. The ingredients list shown in Figure 63.1 came from a box of cheese crackers.

Can you find the words *partially hydrogenated soybean oil* on that list? These words tell you *trans* fats are present. Even though trans fats are in so many foods, most Americans have never heard of trans fat, and have no idea that it may be putting them in danger. Nutrition labels on all packaged foods, like the one shown in Figure 63.2, list the amount of total fat and saturated fat—but say nothing about trans fats.

So, what are trans fats? And if they are not even listed on nutrition labels, then why should you be concerned?

KNOW THE FATS

We eat fat every day as butter, in meat, and in fried and baked foods, but have you ever considered what fat actually is? Fats are made up of special types of molecules called *triglycerides* because they contain three fatty acids and a glycerol chemically bonded together. The fatty acid part determines whether or not it will be listed as saturated fat on your nutrition label.

Look at Figure 63.3, and see that fatty acids have long tails made up of carbon atoms, linked together like a chain by chemical bonds. Those chemical bonds can be single bonds or double bonds because each carbon atom can form a

INGREDIENTS: ENRICHED FLOUR (WHEAT FLOUR, NIACIN, REDUCED IRON, THIAMINE MONONITRATE {VITAMIN B1}, RIBOFLAVIN {VITAMIN B2}, FOLIC ACID), PARTIALLY HYDROGENATED SOYBEAN AND/OR COTTON SEED OIL AND/OR LIQUID SOYBEAN OIL, WHEY, SUGAR, HIGH FRUCTISE CORN SYRUP, BUTTER (CREAM, SALT, ANNATTO COLOR), CHEDDAR CHEESE (MADE FROM CULTURED MILK, SALT AND ENZYMES), BUTTERMILK SOLIDS, SALT, LEAVENING (BAKING SODA, CALCIUM PHOSPHATE), DISODIUM PHOSPHATE, NATURAL FLAVOR, SOY LECITHIN (EMULSIFIER), MATODEXTRIN, ARTIFICIAL COLOR (INCLUDES YELLOW 6), MODIFIED CORNSTARCH, MALTED BARLEY FLOUR, PEANUTS.

FIGURE 63.1   Ingredients on a box of cheese crackers.

total of four bonds. If it forms one bond (a single bond) with each carbon atom on either side of it, it will use its two remaining bonds for two hydrogen atoms. But if it forms two bonds (a double bond) with one carbon molecule and one bond with the carbon on the other side, it will have formed three bonds in total, so it can only bond to one hydrogen atom. Examine the double bond in Figure 63.3 and note that, unlike a single carbon bond, there is only one hydrogen atom (shown as H) linked to each carbon atom.

Nutrition Facts

Serving Size 12 Sandwiches (29g)
Servings Per Container About 9

Amount Per Serving

Calories 150   Calories from Fat 80

	% Daily Value*
Total Fat 9g	**14%**
Saturated Fat 2g	**11%**
Cholesterol less than 5mg	**1%**
Sodium 260mg	**11%**
Total Carbohydrate 16g	**5%**
Dietary Fiber 0g	**0%**
Sugars 4g	
Protein 2g	

FIGURE 63.2   Typical nutrition label.

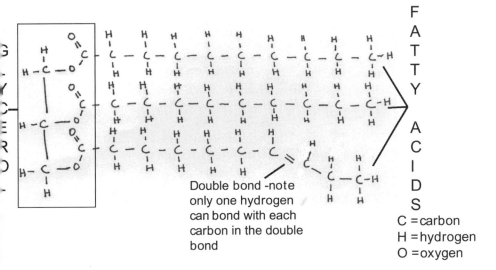

Double bond -note
only one hydrogen
can bond with each
carbon in the double
bond

F
A
T
T
Y

A
C
I
D
S

C = carbon
H = hydrogen
O = oxygen

IGURE 63.3   Partial structure of a triglyceride.

TRIGLYCERIDE: A FANCY WAY OF SAYING *FAT* OR *OIL*

If the fatty acid chains of a triglyceride contain only single-carbon bonds, they will have bonded to the maximum number of hydrogen atoms. Since this fat is filled up with hydrogen atoms, like a sponge filled up with water, it would be listed as *saturated fat* on your nutrition label. Can you find the saturated fat on the nutrition label shown in Figure 63.2? Saturated fat mainly comes from animals. Dairy products such as butter and yogurt contain saturated fat. Saturated fats have a straight shape, so the molecules can pack closely together to form a solid. That is why saturated fats, such as butter and lard, are solid instead of liquid at room temperature. Oils, on the other hand, come mostly from plants. They are also triglycerides, but instead of having only single bonds, they have double bonds between some of the carbons in their fatty acid chains. Since they do not contain the maximum number of hydrogen atoms, like a sponge that could still soak up more water, they are called *unsaturated fats*. The double bonds in plant oils cause their fatty acid chains to bend, which prevents them from packing closely together. That is why unsaturated fats, or oils, are liquid at room temperature.

PLAYING WITH FAT

If saturated fats have only single bonds and unsaturated fats have some double bonds, then what are these mysterious trans fats? *Trans fats* are vegetable oils that were once unsaturated, as all vegetable oils are in their natural state, but were then changed by an industrial process known as *hydrogenation*. This process breaks double bonds in the unsaturated oil and adds hydrogen atoms in place of those double bonds. To hydrogenate oil, hydrogen gas must be bubbled through a mixture of hot vegetable oil and powdered metal. The heat and metal cause the double bonds in the oil to break, leaving single bonds and freeing up more spaces for hydrogen to attach. Remember that double bonds cause fatty acid chains to bend, which prevents them from packing close together. After hydrogenation, when double bonds are removed and replaced by hydrogen atoms, the fatty acid chains become straighter and are able to pack closely together. The result is a solid rather than a liquid. Hydrogenation makes unsaturated oils act more like saturated fats. So, where does the trans come in?

Hydrogenation may add a lot of hydrogen to oil, but it does not get rid of all the double bonds. Here is the problem: *hydrogenation changes the way hydrogen atoms are arranged around the remaining double bonds.* Sound complicated? It's actually quite simple. Normal, natural oils have their hydrogen atoms arranged around a double bond in what is called a *cis* configuration, where the hydrogen atoms are on the *same side* of the double bond (see Figure 63.4).

Hydrogenation changes that cis configuration to a trans configuration, where the hydrogen atoms are on *opposite sides* of the double bond. Oils that have been hydrogenated are called *trans fats* because their double bonds have this unique trans arrangement. The next time you find yourself in the cookie section of the supermarket, look at some cookie and cracker boxes to see if any of them have "No Trans Fats" written on the front.

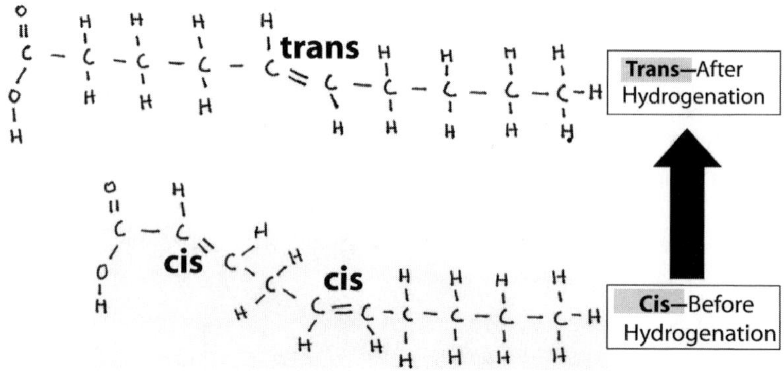

FIGURE 63.4 Comparison of *trans* and *cis* double bonds.

WHY HYDROGENATE?

Now that you know what a trans fat is, you may be wondering why people go through all that trouble making it. And why are they putting it in your food? The answer is simple: it saves food companies a lot of money. Hydrogenation turns unsaturated oils into more saturated fats, which are straighter and more "packable." Since trans fat molecules are packed closely together, like saturated fats, they have a higher melting point than oils. At warm temperatures, they will remain solid and won't spoil as fast as oils do. Food companies add trans fats to their products to make them more stable, that is, to keep food from spoiling. Food that contains trans fats can sit on the shelf at the supermarket for a longer amount of time. If it spoiled, it would have to be thrown away, and the food companies and supermarkets would lose money. So why don't the food companies just use fat that is already saturated, like butter, instead of hydrogenated vegetable oil (trans fat)? Saturated fats, which are mainly animal fats, are expensive. Vegetable oils, on the other hand, are much cheaper. That's right: food companies add trans fats to your food because it saves them a lot of money, even though research shows that they are bad for you.

Hydrogenation is not new—it has been around for over 100 years. In 1897, a French chemist named Paul Sabatier invented the process by turning liquid vegetable oil into margarine. By 1909, food producers in America began hydrogenating cottonseed oils to make trans fats that kept their food from spoiling or melting on hot days. This discovery proved to be so helpful in preserving food that it won Sabatier the Nobel Prize for Chemistry in 1912! After World War II, new metal catalysts were discovered that allowed people to hydrogenate corn, soybean, and canola oils, which contain three double bonds. Then, in the 1970s, trans fats got another boost when new research showed that dietary saturated fat increased the risk for heart disease. Since butter is a saturated fat, many people stopped eating it and switched to margarine, which is hydrogenated but less saturated than butter. What those people did not know was that some margarines contained large amounts of trans fats. Should they have been concerned?

HEARTBROKEN

Scientists began investigating the effects of trans fats on human health in the 1970s. Since then, several large-scale population studies have linked trans fats to heart disease. Many scientists believe that the Nurses Health Study conducted by Frank Hu, Meir Stampfer, JoAnn Manson, Eric Rimm, Graham Colditz, Bernard Rosner, Charles H. Henneken, and Walter Willett (1997) provides the best evidence for this link because it followed more than 80,000 women for fourteen years and took into consideration many dietary and lifestyle factors that could have made the results less accurate. This study found that women who consumed the most trans fatty acids had the greatest risk for heart disease. That means that they had more heart attacks and died more often from heart attacks than women who consumed lower amounts of trans fatty acids. The study showed that women who replaced just 2 percent of their fat intake from trans fatty acids with cis fatty acids found in vegetable oils decreased their risk of suffering and dying from heart attacks by 53 percent! Research by Alberto Ascherio and Walter Willett (1997) has shown that if we replaced trans fats with natural vegetable oils in our diets, we could prevent 30,000 to 100,000 premature coronary heart disease deaths each year. Another study that provided strong evidence for the effects of trans fats on our health is the Alpha-Tocopherol Beta-Carotene Study (ATBC). This study, conducted by a large group led by Pirjo Pietinen (Pietinen et al. 1997), indicated that men who consumed high amounts of trans fatty acids were more likely to die from heart disease than men who consumed smaller amounts. Other studies, such as that conducted by Ronald Mensink and Martijn Katan (1990), have shown that trans fats increase LDL ("bad") cholesterol and decrease HDL ("good") cholesterol. The effect of trans fatty acids on the ratio of "bad" to "good" cholesterol has been shown to be double the effect of saturated fat! A high ratio of LDL to HDL cholesterol increases the risk for heart disease.

TRANSCENDING NATURE

By now, you may be convinced that trans fats are bad for you. But you still probably do not know what makes those funny little fatty acids so bad for your heart. And guess what: neither do the rest of us! That's right. Scientists still have not figured out just how those trans fats weave themselves into your body and increase your risk for heart attacks. But some people have come up with their own theories for why trans fats and humans do not mix well. For one thing, most trans fats do not exist in nature—they are created by people through the process of hydrogenation. The only trans fatty acids that exist in nature are made by bacteria in animals such as cows and sheep, which have a special type of stomach called a *rumen*. These bacteria only produce very small amounts of trans fat. For millions of years, no animals—not even human ancestors—ate foods with trans fats. Some people believe that because we did not evolve eating trans fats, our bodies do not know how to use and process them.

If you plug the term *trans fats* into a search engine on the Internet, you will find hundreds of websites blaming them for everything from blindness to cancer. While scientists have not linked trans fats to any condition other than heart disease, many people are concerned because fatty acids make up a major part of the membrane that surrounds every single cell in your body. Unsaturated fatty acids are said to be the "chemically active" parts of cell membranes because, as you already know, unsaturated fatty acids contain double bonds. Reactions usually take place at double bonds because they contain two pairs of shared electrons, one of which is eager to bond to other molecules. In cell membranes, the double bonds in fatty acid chains bind to proteins and other molecules passing through. Those double bonds let in important molecules that the cell needs to survive, such as oxygen, and keep bad molecules out. But if a trans fatty acid somehow got put into a cell membrane instead of the normal cis fatty acid, could it still do its job? Remember, natural unsaturated fats occur in a cis configuration, with hydrogen atoms on the same side of the double bond. That cis arrangement causes the fatty acid to bend at the double bond, so that the double bond sticks out, ready to grab onto any passing molecule that needs an electron. Trans fats, on the other hand, have hydrogen atoms on opposite sides of the double bond. That trans arrangement keeps the fatty acid straight, instead of bent, so the double bond stays tucked within the hydrogen atoms on the fatty acid chain. Some people think that the shape of trans fats makes them less able to latch onto passing molecules, such as oxygen, that cells need to survive. Or, they could allow toxic molecules into the cell by not latching on when they are supposed to. What do you think?

TRANS ACTION

If you start paying attention to the ingredients in the foods you eat, you will soon realize that trans fats have found their way into just about every packaged food, from breath mints to peanut butter. Because of the known and unknown health effects of trans fats, if they were introduced today, the government would not allow them in our food! So, what is being done about trans fats? Since 2006 the U.S. Food and Drug Administration (2006) requires all food producers to list the amount of trans fats directly under the amount of saturated fats on their nutrition labels. That does not mean that trans fats will stop showing up in your food, but at least it will be easier to find them.

What can you do to protect yourself? When you feel like you have just got to have those French fries, settle for a smaller portion. Shop in health food stores—they usually refuse to carry any food that contains trans fats. Do not settle for food made with heart-damaging ingredients just so that they can sit on the shelf longer. A recent study on rats reported by Helen Phillips (2004) showed that eating too much trans fats can affect learning and memory, and even damage the brain! Minimize your consumption of trans fats—we are just beginning to understand the effects that they may have on our health.

REFERENCES

Ascherio, Alberto, and Walter Willett. 1997. Health effects of trans fatty acids. *The American Journal of Clinical Nutrition* 66:1006S–1010S.

Hu, Frank B., Meir J. Stampfer, JoAnn E. Manson, Eric Rimm, Graham A. Colditz, Bernard Rosner, Charles H. Henneken, and Walter Willett. 1997. Dietary Fat Intake and the Risk of Coronary Heart Disease in Women. *New England Journal of Medicine* 337:1491–99.

Le Couteur, Penny, and Jay Burreson. 2003. *Napoleon's Buttons: How 17 Molecules Changed History.* New York: Tarcher/Putnam.

Mensink, Ronald P. M., and Martijn B. Katan. 1990. Effect of Dietary *Trans* Fatty Acids on High-Density and Low-Density Lipoprotein Cholesterol Levels in Healthy Subjects. *New England Journal of Medicine* 323:439–45.

Phillips, Helen. 2004. Fears Raised over the Safety of Trans Fats. *New Scientist* 184 (2472): 17.

Pietinen, Pirjo, Alberto Ascherio, Pasi Korhonen, Ann Hartman, Walter C. Willett, Demetrius Albanes, and Jarmo Virtamo. 1997. Intake of Fatty Acids and Risk of Coronary Heart Disease in a Cohort of Finnish Men: The ATBC Study. *American Journal of Epidemiology* 145:876–87.

U.S. Food and Drug Administration. 2006. *Trans Fat Now Listed With Saturated Fat and Cholesterol on the Nutrition Facts Label.* www.cfsan.fda.gov/~dms/transfat .html.

64

Hair Relaxers: The Chemistry of an Everyday Experience

Marsha Smith and Catherine Milne

INTRODUCTION

Science is everywhere. On a daily basis, millions of chemical reactions take place in the cells of your body. Evidence for these reactions comes from the things you can observe such as the heat of your skin, the materials you excrete, and the food you have to eat to keep going. Like all of us, you are actively involved in chemistry. But how aware are you of the chemistry behind the activities with which you are involved? Indeed, many of the things you use to make your life easier or more interesting only work because of chemical reactions. Even though they might not be the first objects that you think of, *chemical hair relaxers* are a case in point.

Although the history behind the origin of chemical relaxers is not well known, many African American women are grateful for their creation. African Americans are blessed with the precious gift of having overly curly hair. However, given the opportunity to make it straight, 90 percent of African American women have relaxed their hair at some stage in their lives. In some families relaxing of the hair is a "rite of passage," because it gives the adolescent or preteen a feeling of maturity and freedom. However, despite the different reasons for people deciding to change the structure of their hair, few ever consider the chemical transformations that occur as overly curly hair is transformed into straight hair in a process called *relaxing*. Relaxing the hair, which allows women of African American descent (for example) a wide range of varying hairstyles, requires the use of chemicals called *hair relaxers*. Women walk into the beauty salon and allow perfect strangers to touch, feel, and apply relaxers to their hair, but how many of them understand the science behind the process? In order to understand how hair relaxers work, you need to know a little about the chemical composition of hair, the structure of hair, and the chemicals involved in this process.

CHEMICAL COMPOSITION OF HAIR

Human hair is composed of a root and a shaft. The hair shaft, which is the part of the hair that you see above the skin, is composed of a type of hardened protein called *keratin*. Keratin is a structural protein and has a different structure from the globular proteins such as enzymes with which you might be more familiar. However, like all proteins, its building blocks are composed of five chemical elements: carbon, oxygen, nitrogen, hydrogen, and sulfur. Different colored hair contains different combinations of these elements. Dark hair usually contains the highest percentage of carbon and the lowest percentage of sulfur, and light hair less carbon and hydrogen and more oxygen and sulfur. Amazing to think that the combination of these elements, which accounts for hair color, is based on a combination of genes inherited from our parents.

CHEMICAL BONDING IN HAIR

The building blocks that make up keratin are smaller molecules called *amino acids*. Three amino acids chemically linked are called a *peptide*. Peptides are arranged in long lengths called *polypeptide chains*. In order to form these chains, the amino acids are joined together by *peptide bonds*, and the correct number of amino acids placed in their correct order will form a specific protein, for example keratin (see Figure 64.1 for a schematic diagram). Peptide bonds are strong covalent bonds that form between a nitrogen atom on one amino acid and a carbon atom on another. Peptide bonds are the bonds that help form the "skeleton" of the protein. Strands of the protein are long and thin, and they are arranged side by side to form fibers. The strength of the fibers comes from them being twisted together like rope.

Have you ever wondered why hair does not dissolve when you wash it? Your skin is also largely made up of criss-cross fibers of keratin helping to make your skin water-resistant and the first line of defense against infection. Or have you wondered why your hair is longer when it is wet than when it is dry? The answers to these questions lie in the structure of keratin and the way in which the polypeptide chains of keratin interact chemically with each other.

Structural proteins, like keratin, tend to be made up of polypeptide chains composed of a small number of different amino acids, unlike globular proteins that tend to have many more different types of amino acids in their polypeptide chains. In keratin one of the important amino acids is called *cysteine*. As the polypeptide chains are formed, various chemical groups interact electrolytically to form hydrogen bonds between adjacent polypeptide chains that assist the coiling of the protein into fibers. These fibers are strengthened by covalent bonds between cysteine amino acids between adjacent polypeptides (see Figure 64.1). Because the bond is between two sulfur atoms, one from each of the cysteine amino acids, these covalent bonds are called *disulfide bonds*. These are very strong bonds and, along with the water-hating quality of some of the amino acids present in keratin, contribute to making the protein insoluble in water.

Hydrogen bonds are responsible for the hair's elasticity and flexibility. When your hair is wet, water provides a ready source of oxygen atoms needed for hydrogen bonding, but as the hair dries and the water evaporates, the hydrogen bonds between O-H groups on different polypeptide chains reform, helping to put the coil back in your hair and making the hair strands shorter. You have more hydrogen bonds in your hair when it is dry than when it is wet. Some people, perhaps those without curly hair, make use of hydrogen bonds when they put their wet hair in curlers or rollers. As the hair dries, the hydrogen bonds in the hair form and help to hold the curls in the hair.

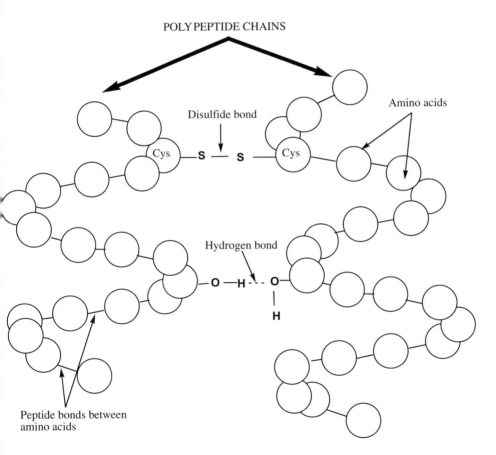

FIGURE 64.1   Schematic diagram showing the position of peptide bonds and disulfide bonds in two polypeptide chains of a generalized structural protein molecule like keratin.

According to John Gray (2003), even the Egyptians used curlers and the hair's capacity to form hydrogen bonds. At the time of the pharaohs, they applied mud and wooden rollers to the hair. Once the hair was dry, the mud was shaken out. One can only guess at the fashion statement that resulted. However, whether you are an ancient Egyptian or a modern city dweller, as soon as the hair is wet or if the air is humid, the hair will lose its curl as the proteins form hydrogen bonds with the water in the environment.

Peptide and disulfide bonds are covalent bonds. A covalent bond is formed by a sharing of a pair of electrons between two atoms of a molecule, forming the strongest of all chemical bonds. Covalent bonds tend to be at least ten times stronger than hydrogen bonds. As the name implies, hydrogen bonds depend on electrolytic differences between hydrogen atoms and other atoms, most commonly oxygen. These bonds form between molecules. In proteins, the protein-protein hydrogen bond is weaker than the protein-water hydrogen bond, which explains the greater affinity of keratin for water when your hair is wet. Ionic bonds are also electrolytic and require the transfer of electrons from one atom to another, resulting in the formation of positively and negatively charged ions within a compound structure. In the presence of water, ionic bonds tend to be much weaker than covalent bonds.

HAIR LAYERS

As mentioned above, hair is composed of a root and a shaft. The shaft is the focus of interest in the chemistry of hair relaxers. What you call *hair* is largely composed of keratin. As three polypeptide chains of keratin coil together, they form a structure called a *protofibril*. These protofibrils are packed together in an eleven-stranded cable called a *microfibril*, and hundreds of microfibrils are cemented into a bundle called a *macrofibril*. So as you can see there are a lot of polypeptides in a macrofibril. Macrofibrils are grouped together to form a layer of hair called the *cortex*. Hair is composed of three concentric layers: *cuticle*, cortex, and *medulla*. The cortex, the middle layer, also contains melanin, a chemical that gives hair its color. The medulla, the innermost layer, is really a tube. The cuticle, the outer layer of the hair, consists of packed dead cells or scales and surrounds the cortex. The cells in the cortex are more sensitive to the effects of alkalis and other chemicals than the cells of the cuticle. However, the cuticle often provides the most obvious signs of hair health because the shine associated with healthy hair is due to close packing of scales, and if the scales start to lift at the edges the hair begins to lose its shine. In both the cortex and the cuticle, keratin is the major structural molecule.

Since breaking hydrogen bonds does not have a permanent effect on your hair, if you want to permanently change the shape of your curly or straight hair, you need to use a chemical reaction that will "break" the disulfide bonds. For African American women, applying strong alkali chemicals, such as guanine hydroxide or sodium hydroxide, to the hair provides the necessary elements that will cause the "breaking" of the disulfide bonds, allowing them to "relax" their hair. It is the breaking of these bonds that is so important to the chemistry of hair relaxers.

WHAT CHEMICALS ARE INVOLVED IN HAIR RELAXING?

According to Karen Shelton (2003), the basic products that are used during chemical hair relaxing are (1) a chemical hair relaxer, (2) a neutralizer, (3) a protein-rich moisturizer to stabilize the hair, and (4) a petroleum cream, which is used as a protective base to protect a client's scalp during the chemical straightening process. Hair has a pH of 4.5–5.5, indicating that it is acidic. On the other hand, all relaxers are alkalis. Guanidine hydroxide, a no-lye relaxer (lye is the common name for sodium hydroxide and in the past was extracted from the ash of woodfires), has a pH of 9–9.5, and sodium hydroxide has a pH of 12–14. All relaxer treatments leave the hair in an alkaline state with a pH of greater than 7. To rebalance the hair to its natural acidic pH level, acidic "normalizing" or "neutralizing" shampoos with a pH of 4.5–5 are usually used after the cream relaxer is rinsed from the hair.

SO WHAT IS CHEMICAL HAIR RELAXING?

Chemical hair relaxing is the process of permanently rearranging the basic structure of overly curly hair into a straight form. When done professionally, it leaves the hair straight and in a satisfactory condition, to be set into almost any style. Chemical hair relaxing involves three basic steps: processing, neutralizing, and conditioning.

Processing

During the processing step, the chemical relaxer is applied and the hair begins to soften so the chemical can penetrate to loosen and relax the natural curl. So what is happening chemically to your hair? The alkali lotion causes the scales of the cuticle to open slightly, which allows the alkali to flow into the cortex. As the solution penetrates into the cortical layer, the OH⁻ from the sodium hydroxide strips a hydrogen atom from the carbon second closest to the sulfur in the disulfide bond and initiates a series of chemical reactions. These chemical reactions involve the breaking of disulfide bonds and the formation of cysteine that lacks these bonds (see Figure 64.1). The polypeptide chains can then move apart and slide against each other, which can be observed as the hair seems to soften and "fluff up."

Neutralizing and Conditioning

As soon as the hair has been sufficiently straightened, the chemical relaxer is thoroughly rinsed out with warm water, followed by either a built-in shampoo neutralizer or a prescribed shampoo and neutralizer. The neutralizer, which commonly contains oxidizers such as hydrogen peroxide, assists the reforming of the disulfide bonds and consequent hardening of the hair into a straight shape. As the neutralizer reforms the cysteine/disulfide cross-bonds into their new position and rehardens the hair, the hair presents a straightened structure. After the neutralizing step, the hair must be conditioned. Depending on a person's needs, the conditioner may be part of a series of hair treatments, or it may be applied to the hair after the relaxing treatment.

You might also note a similarity between using hair relaxers for overly curly hair and using "perms" for straight hair. A permanent wave is achieved also by using similar chemicals to break and reform the disulfide bonds. First, the hair is washed and wound onto formers so that after the application of the alkali and neutralizer, the hair will be hardened into a curly shape.

Although some people do choose to straighten their own hair at home, to prevent damage and burns to the skin and scalp, it is strongly recommended that a professional with a prior record of straightening experience perform this process. Once hair has been chemically straightened, it requires extra special care to overcome the effects of the drying chemicals. Relaxed hair will require deep conditioning treatments at least one or two times a week. While straightening offers a new range of hairstyles to many women, it should not be undertaken lightly because it may damage the hair and requires from each woman who decides to relax her hair an ongoing commitment to postrelaxing conditioning and special hair care, a decision not to be taken lightly. This advice is also appropriate for those considering "perming" their hair.

CONCLUSION

Understanding the chemistry of hair relaxers requires an understanding of the relationships between the submicroscopic world of protein structure, the bonds that hold the structure in place, and chemical relaxers. The cortex layer in the hair is composed of long flat scales that contain molecules of the large structural protein, keratin. Within keratin, atoms in the amino acids are linked together by various bonds, the most important of which for hair relaxing are hydrogen and disulfide bonds.

The disulfide bonds are located between two cysteine amino acids on separate polypeptide chains within the keratin protein. Chemically, disulfide bonds are much stronger than hydrogen bonds. Consequently, a disulfide bond can only be broken by a chemical reaction involving acidic hair molecules and a strong alkali relaxer such as sodium hydroxide, not by heat or water, which is all that is necessary to break hydrogen bonds. If used properly, chemical hair relaxers do not alter the primary structure of the protein because they do not break the peptide bonds linking together the amino acids in individual polypeptide chains of keratin. When a chemical relaxer is applied to the hair, the disulfide bonds are broken by a strong alkali and then reformed by the application of a neutralizer. However, the new disulfide bonds do not reform exactly where they were formed before the relaxer was applied; therefore, chemical relaxing permanently rearranges the structure of the hair. The changes can be observed because the hair is now straight when prior to the treatment it was curly.

Understanding chemical bonding can provide a richer insight into what is happening to your hair at the molecular level when you wash it, dry it, set it in rollers, relax it, or perm it. Understanding the role of chemical bonds in the behavior of your hair might also lead you to understand why it is easier to get crumbs off the plate with a wet finger rather than a dry one and why you are more likely to leave fingerprints behind when your hands are moist rather than when they are dry.

REFERENCES

Gray, John. 2003. *The World of Hair: An Online Reference*. P&G Hair Care Research Center. www.pg.com/science/haircare/.

Shelton, Karen. 2003. Chemical Hair Straightening FAQs. October. www.hairboutique .com/tips/tip086.htm.

65

Sunscreens and Sunblocks: Using Sunprint Paper to Investigate SPF

Ji-Myung Nam and Catherine Milne

WHAT ARE SUNBLOCKS AND SUNSCREENS?

Sunblocks and sunscreens protect us from the hazardous UV rays in sunlight in different ways. Sunblock reflects sunlight away from our skin and prevents damaging UV rays from hitting our skin. The most popular sunblock chemicals used in the United States are zinc oxide and titanium oxide, which are white powders added to lotions and creams. Sunscreens function differently and contain chemicals that are designed to act like melanin in our skin by dissipating the energy from hazardous sunlight into heat rather than into light radiation or photoproducts that can cause skin damage. Oxybenzene, octinozate, and padimate are most popular chemicals used in sunscreen products in the United States.

WHAT IS SPF?

SPF (sun protection factor) was developed to be a measure of the effectiveness of sunscreening. The higher the SPF, the more protection a sunscreen offers against UVB (ultraviolet B type radiation), which causes sunburn. The makers of skin care products use a formula to calculate SPF. For example, SPF 15 means that, normally, a user can remain in the sun fifteen times longer than would otherwise cause him or her to burn.

During our daily routine, some of us may put on sunblock and sunscreen to avoid the hazardous effects of the sun. Many believe that sunblock can essentially "block" the sun's harmful rays and protect us from its dangers. Manufacturers claim that their sunscreen and sunblock products protect against skin cancer and reduce skin aging. But, have you ever wondered whether you are putting on enough sunblock or sunscreen to do this? Or whether your sunblock or sunscreen is doing what the manufacturers claim it does? *Sunprint paper* is a tool we can use to discover how effectively sunblock and sunscreen products work on our skin.

The experiment described below was developed to address the following National Science Content Standards, especially at the middle school and high school level:

Content Standard A—Understandings about Scientific Inquiry. You might notice the importance of questions to this investigation. Questions are important to the process of scientific inquiry and determine the structure of the associated investigation. Sometimes it is not possible to answer the question that you want to ask. So a more focused question has to be asked that can be tested. Also, appropriate tools and techniques help you to gather, analyze, and interpret data collected from the investigation.

Content Standard B—Properties of Matter. You might want to think how the active ingredients in sunscreens mimic how melanin works in our skin.

Content Standard F—Personal Health. Many environmental health factors such as ultraviolet (UV) light (usually from the sun) can affect your health.

HOW CAN YOU TEST SUNBLOCKS AND SUNSCREENS?

1. Collect several different kinds of sunblock or sunscreen products (lotion, cream, lip balm, etc.) that you might have in your home. Your test products can include lotions, creams, and lip balms. Search through the different kinds of sunblock and sunscreen for the SPF numbers on them (e.g., SPF 50, 30, and 15).

   What do you think the term *SPF* means? Suppose that without sunscreen or sunblock, you can stay in the sun for ten minutes before starting to burn. When using sunscreen or sunblock, you can stay in the sun your initial ten minutes multiplied by the SPF rating. But what is missing with this definition of SPF? How thickly do I need to apply the sunscreen or sunblock that I have? Will the thinnest scrape do, or do I need to put it on thickly like I put jelly on a piece of bread?

What You Need

- Sunprint paper. This is UV light–sensitive paper that is coated with a photosensitive chemical, so when the paper is exposed to sunlight, it turns blue and the unexposed part turns white. Sunprint paper can be purchased inexpensively at various science museum gift shops or directly from online shops such as Sciencestuff.com (2006) and SacToys.com (n.d.).

  Sunblock and sunscreen creams, lotions, and lip balms, or some other type of sunblock or sunscreen product.

Water, sunlight, butter knife, or plastic knife.

2. To find the active ingredients (i.e., those that actually protect us against the sun), check the ingredients on the package for the use of sunblock or sunscreen chemicals. (Sunblock commonly contains titanium dioxide or zinc oxide. Sunscreens usually contain chemicals such as oxybenzon, octinixate, and padimate.) Look for the active ingredients in the samples you have collected, and make a note of the name and amount of each.

3. Cut the sunprint paper into four (4) pieces, each measuring 6 inches by 4 inches. Select four different kinds of sunblock or sunscreen products that use different kinds of sunblock or sunscreen chemicals to use for your tests (e.g., zinc oxide and padimate). For each piece of the sunprint paper, apply the sunblock or sunscreen product at two different thicknesses: about 1 mm thick, or the thickness of a penny, on one side of the sunprint paper; and on the other side, apply it as though you were applying it to your skin.

4. If the sunblock or sunscreen product is effective, what do you predict will happen to the sunprint paper under each thickness? (Note: At this step, when directly exposed to the sun, sunprint paper becomes lighter and lighter as it reacts with light from the sun.)

5. Test your prediction by exposing the sunprint papers to the sun for 5–7 minutes until the sunprint paper turns almost white.

6. Rinse the sunprint papers with water for about one minute, and wash away the sunscreen and sunblock products. Note that as you wash the papers in water, the parts of the sunprint paper that were exposed to the sunlight turn blue, and unexposed or protected parts turn white!

7. Dry the papers.

8. Examine the color of the sunprint paper on which you had placed the sunblock or sunscreen lotion or cream. Do you notice any differences between the color of the sunprint paper under the sunblock or sunscreen and the sunprint paper that was exposed to the UV rays of the sun? Do you notice any differences in the sunprint paper between the thick and thin layers of sunblock or sunscreen lotion? Was your prediction supported? Which thickness was more effective?

If you tested more than two sunblock or sunscreen lotions or creams with different SPF numbers, did SPF factor make a difference? Can you think of an explanation for any of the differences you observed? What advice would you give to someone using sunblock or sunscreen? Can you think of other questions that you could investigate using sunscreen paper? How might you use clothing and sunglasses?

FACTS ABOUT SUNBLOCKS AND SUNSCREENS

If we want to use sunblocks and sunscreens effectively, we have to put them on our skin as thickly as possible. But most people apply sunblock very thinly (below the effective 1 mm thickness). In that case, the skin cannot be effectively protected by the sunblock or sunscreen. So when you have to stay outdoors in the sun, try to apply sunblock thickly, and wear a hat and sun-protective clothing or use an umbrella to block out the sun.

REFERENCES

SacToys.com. N.d. SacToys. www.sactoys.com.
Science Stuff. 2006. Science Stuff. www.sciencestuff.com.

66

Hydrogen Fuel Cells: The Alternative Energy Source of the Future?

Sophie Homan and Catherine Milne

How does the hydrogen fuel cell work? How can the production prove to be safe, environmentally conscious, and economically sound? Consumer preferences control the choices made in energy markets, in technology development, and in public policy. Education will help the hydrogen fuel cell to gain public acceptance for hydrogen-related products and services.

WHAT WAS THE REAL CAUSE OF THE *HINDENBURG* AIRCRAFT EXPLOSION?

In 1937, the *Hindenburg* exploded while attempting to land outside of Lakehurst, New Jersey. The common assumption is that hydrogen ignited the fire. According to Lee Krystek (2001), a NASA scientist who examined the information available, the initial fire was not burning hydrogen. The doping solution, a liquid applied to stretch the material over the hull and make the material waterproof, was a compound containing a layer of iron oxide, cellulose butyrate acetate, and powdered aluminum—the equivalent of rocket fuel! Because the airship was flying right through an electrical storm, all that was needed was a single spark to set the fire.

People's concern about the flammability of hydrogen leads them to be concerned about the safety of hydrogen fuel. However, every fuel is flammable, and if not stored appropriately, all fuels are dangerous. The development of the hydrogen fuel cell provided a method for harnessing the power of hydrogen. Fuel cells use combustion of hydrogen to produce electricity, but the fuel does not burn because the half-reactions that are together in combustion reactions are separated in a fuel cell and because the electrons are transferred through an external circuit. Any fuel containing hydrogen, such as methane and natural gas, can be used, but only hydrogen gas can be used directly. All of the fuel cell systems and hydrogen storage techniques are engineered with safety being the primary concern. The tanks used to store liquid and gaseous hydrogen are required to undergo rigorous safety testing before they are certified for hydrogen storage. But how do hydrogen fuel cells work, and do they offer a clean, safe fuel that could be an alternative to coal and gasoline?

WHAT'S THE SCIENCE BEHIND IT ALL?

One of the common types of fuel cells is the Polymer Electrolyte Membrane (PEM) fuel cell. According to the U.S. Department of Energy (2004), the PEM fuel cells consist of an electrolytic membrane sandwiched between an anode and a cathode. Looking like plastic wrap, only thicker, the electrolytic membrane conducts the positively charged hydrogen ions, also known as *protons*, through to the cathode.

The anode is the electrode at which oxidation (or loss of electrons) takes place (see Figure 66.1 for a schematic diagram of a fuel cell). This electrode is negatively charged and is made of graphite with a platinum catalyst, which, like all catalysts, speeds up the oxidation process without interacting with the

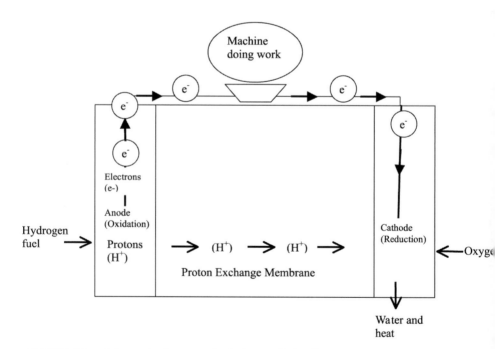

FIGURE 66.1   Schematic diagram of a hydrogen fuel cell.

materials involved. Catalysts play a key role in determining the performance of a fuel cell system. In a fuel cell, hydrogen is sent to the anode, where it is dissociated. If the positively charged hydrogen ions travel through the PEM to the cathode, where do the electrons go? They flow around the membrane through an external circuit. The flow of electrons from one place to another is called *electricity*—which can be used to power machinery such as cars and trucks. The reaction that takes place at the anode is

$$2H_2 \xrightarrow{\text{Pt catalyst}} 4H^+ + 4e^- \quad \text{(Oxidation)}$$

According to this half-reaction that occurs at the anode, two molecules of hydrogen gas produce four hydrogen ions and four electrons. The cathode at the other side of the fuel cell is the electrode at which reduction (gain of electrons) takes place. This electrode is positively charged and is also made of graphite with a platinum catalyst, which speeds up the rate of reduction. Oxygen is sent to the cathode, and combines with the electrons from the external circuit and with the positively charged hydrogen ions from the PEM. This combination produces water and heat, which are the only by-products of the hydrogen fuel cell! The half-reaction that takes place at the cathodes is

$$O_2 + 4H^+ + 4e^- \xrightarrow{\text{Pt catalyst}} 2H_2O \quad \text{(Reduction)}$$

The combination of these two reactions, showing the overall chemical reaction, can be represented as follows:

$$2H_2 + O_2 \rightarrow 2H_2O$$

Note that in this process, the hydrogen is oxidized (it loses electrons) and the oxygen is reduced (it gains electrons). Using hydrogen fuel and oxygen from the surroundings, it is possible to produce energy as heat (not useful) and work (useful because it can be used to power machines), with water as a by-product. So fuel cells rely on a simple spontaneous (it does not require energy to take place, and the chemical reaction releases energy) redox reaction, but making this an efficient controlled reaction that produces useful energy challenges the ingenuity of chemists and engineers.

Flow plates make up the external circuit that provides a pathway for electrons from the anode to the cathode. They also channel hydrogen and oxygen to the electrodes and channel water and heat away from the fuel cell. In order to produce usable levels of energy, fuel cells have to be collected into stacks much like electric cells in a battery.

However, it is important to consider more than the chemistry that happens inside the fuel cell, especially the by-products produced from these chemical reactions. Ideally, the only by-products are hydrogen, oxygen, water, and heat. However, obtaining pure hydrogen fuel presents a challenge. So the parts of the system involved in producing, storing, distributing, and using hydrogen need to be examined.

WHAT ARE THE ENVIRONMENTAL AND ECONOMIC IMPACTS OF HYDROGEN?

Production

Because hydrogen is not a readily available energy source, it is important to identify compounds containing it and then separate out the hydrogen. At this time, several processes are being used. One is the electrolysis of water, which involves putting an electrical charge through water to break up the water molecules into the elements that comprise them: hydrogen and oxygen. Water is a renewable resource, and the only by-products are hydrogen, oxygen, and heat. However, the energy needed to electrolyze water to produce oxygen and hydrogen is currently obtained by processing a variety of fossil fuels, which in turn produce by-products harmful to the ecosystem. This creates a conundrum. It actually costs less to produce power from fossil fuels than it does to produce power from hydrogen fuel cells if electrolysis is used to produce the hydrogen for the fuel cells in the first place. The one advantage of using water as the source of hydrogen is that it provides an option for the production of electricity from renewable resources. However, a question that arises is "Do fuel cells produce more energy than the energy required to produce the hydrogen needed for fuel cells?" This remains a vexed question because improving the efficiency of the fuel cells requires more funding for research and development than is currently available.

Another way to obtain hydrogen from natural resources is via the processing of fossil fuels. An example of this is a process called *steam methane reforming*, the reaction of natural gas with steam to produce a mixture of hydrogen and carbon dioxide.

The chemical equation for the steam methane reaction is as follows:

$$\text{Step 1: } CH_4 + H_2O \rightarrow CO + 3H_2$$

$$\text{Step 2: } CO + H_2O \rightarrow CO_2 + H_2$$

This method of using methane (CH_4) produces twice as much hydrogen as electrolysis does, but its by-product, carbon dioxide (CO_2), is a greenhouse gas and therefore has greater negative impact on our environment.

In the immediate future, hydrogen will most likely be produced by steam reforming of natural gas and through electrolysis of water using electricity from conventional energy resources. However, in the long term, alternative resources could be used to provide the initial electricity, such as solar or wind power; and if hydrogen production technologies such as these are developed, it will become more cost-effective to produce hydrogen. Because hydrogen can be produced from a wide variety of resources, each region of the country might be able to use a different combination of resources to produce it.

Distribution and Storage

Operation of hydrogen fuel cells requires a constant supply of hydrogen gas, so distribution and storage also become issues. A reliable and low-cost hydrogen distribution system cannot be built overnight. Limited hydrogen pipeline networks exist in certain regions of the United States to supply it to the refining industry. Gas production plants also frequently transport hydrogen by tanker truck to industrial users. As hydrogen demand grows, industry will respond by building and expanding the hydrogen delivery and distribution network using current and future technologies for pipeline construction, hydrogen storage, and delivery.

Hydrogen has the potential to become a leading energy source, by reducing U.S. dependence on imported petroleum, and by reducing pollution and greenhouse gas emissions. It can be produced in large refineries in industrial areas, in power parks and fueling stations in communities, and in distribution facilities in rural areas, making it a readily available energy source. Streamlined processes can greatly reduce the release of carbon dioxide and other pollutants into the atmosphere.

WHAT ARE THE APPLICATIONS OF HYDROGEN POWER?

Besides being a possible energy source for stationary power, hydrogen has also been researched and trialed for use in transportation. Spacecraft already use hydrogen power (some of the Gemini spacecraft used experimental fuel cells), and researchers are working on the development of the hydrogen automobile. Other uses include portable power for applications such as portable generators, laptop computers, and other handheld electronics. However, these applications need further design development in order to compete in the market. Despite the great benefits, there is still a lack of consumer demand for hydrogen power, perhaps associated with a lack of awareness of hydrogen fuel cells and the current inability of hydrogen fuel cells to compete on the open market with fossil fuels and nuclear fuels, and this may inhibit its success. Part of this problem might be associated with the lower level of funding for fuel cell research and development when compared with the level of funding available to fossil fuels and nuclear fuels research, technology, and development.

WHY HYDROGEN? A CONCLUSION

Once you have a source of pure hydrogen, hydrogen fuel cells offer so many advantages that it is difficult to understand why more people are not eager to use them in a range of applications. Hydrogen emits no pollutants if we create it in an environmentally friendly way through the use of renewable resources and the electrolysis of water. Also, if hydrogen became our primary fuel source, we would not have to spend as much money defending our access to oil overseas. The hydrogen industry could create a variety of new jobs in research and industry. The best thing about a renewable resource is simply that it is renewable. We must stop depending on resources that are limited in supply and look to resources that are in an abundant supply. The chemistry behind the hydrogen fuel cell is relatively simple; however, the issues and problems that arise in trying to implement this technology make matters very complicated. The best way to create a demand for hydrogen fuel cell technology is through education. On a local scale, we must educate those in our communities about the benefits of the hydrogen fuel cell technology; and on a global scale, we must come to realize two things: that hydrogen is a viable alternative to the processing of fossil fuels and that our fossil fuel supply is not renewable.

REFERENCES

Krystek, Lee. 2001. "The Mystery of the Hindenburg Disaster." *Unnatural History Museum.* www.unmuseum.org/hindenburg.htm.

U.S. Department of Energy. 2004. *Hydrogen, Fuel Cells, & Infrastructure Technologies Program: How a Fuel Cell Works.* US Department of Energy—Energy Efficiency and Renewable Energy. www.eere.energy.gov/hydrogenandfuelcells/.

Index

About the Editor and Contributors

Kenneth Tobin is Presidential Professor of Urban Education at the Graduate Center of City University of New York. In 2004 Tobin was recognized by the National Science Foundation as a Distinguished Teaching Scholar and by the Association for the Education of Teachers of Science as Outstanding Science Teacher Educator of the Year. His research interests are focused on the teaching and learning of science in urban schools, which involve mainly African American students living in conditions of poverty. A parallel program of research focuses on coteaching as a way of learning to teach in urban high schools.

Jennifer D. Adams is a manager of professional development at the American Museum of Natural History. She is also a Ph.D. candidate in urban education at the Graduate Center of the City University of New York. Her research interests include informal education, urban science teaching, learning, and learning to teach, and arts education.

Michele Amoroso is an elementary school teacher at the Queens College School for Math, Science, and Technology in New York. She has been teaching in New York City for over ten years and is known for her holistic approach to teaching. Her research interests concern the elements that contribute to classroom culture and ways to ensure the development of positive emotional learning environments. As a former eco-tour leader, she uses her passion for environmental science to teach and study the learning of her early and elementary childhood students.

Angela Calabrese Barton is an associate professor of science education and director of the Urban Science Education Center at Teachers College, Columbia University. Her research interests focus on issues of equity and social justice in science teaching and learning in urban centers, urban science education, and critical/feminist perspectives on science education.

Jennifer Beers is a high school biology teacher at Mastery Charter High School in Philadelphia. Jen is known for her use of innovative and culturally adaptive teaching practices, as well as her application of cogenerative dialogues to improve the teaching and learning of science. She has worked extensively with university researchers like Sarah-Kate LaVan and has published several articles on her experiences as an urban science teacher. Currently, Jen has returned to school to seek an additional certification in special education.

Foram Bhukhanwala is a doctoral student and graduate teaching assistant in the Department of Elementary and Social Studies Education at the University of Georgia. She completed her B.S. and M.S. degrees in human development at the University of Mumbai, India, where she was involved with public policy and health issues for young children. Foram has experiences working in India with children on the street and with preservice and in-service teachers. Her research interests center on issues of social justice and equity inherent in democratic education. In particular, she is interested in arts-based experiences as a tool for working with people who are oppressed.

William J. Boone is Condit Professor of Science Education at Miami University, Ohio. For over a decade, he has conducted research with respect to the issue of evaluation in science education. He holds a Ph.D. from the University of Chicago's Program in Measurement, Evaluation and Statistical Analysis, Department of Education. His research interests are the development and analysis of survey and test data. He has evaluated data for a wide variety of stakeholders (e.g., schools and museums).

Lisa M. Bouillion is on the faculty of the College of Education at the University of Illinois, Urbana-Champaign and is also the state specialist in youth development and technology innovation for the UIUC Office of Extension. Bouillion has been involved in several national school reform efforts, including the New American Schools Project and U.S. Department of Education Technology Literacy Initiative. Her research interests are focused on understanding how to leverage home-school-community partnerships to create technology literacy experiences for youth and adults, particularly in low-income and minority communities.

David J. Brady is the Addy Family Professor of Electrical and Computer Engineering at Duke University. Professor Brady has published over seventy-five refereed publications in optical sensing and information systems, is the founding scientist of Centice Corporation and Distant Focus Corporation, and is the founding director of the Fitzpatrick Institute for Photonics at Duke. Professor Brady has also long been involved in initiatives in technical education, mostly focused on undergraduate and graduate engineering but also occasionally in support of Margery Osborne's research.

Shannon Casey teaches seventh-grade life science at Wagner Middle School in New York City and teaches research and writing skills to high school students enrolled in New York University School of Medicine's High School Fellows Program. She received her B.S. in biological anthropology and human biology from Emory University and her M.A. in science education from New York University. Prior to teaching, she was the manager of youth services at New York Cares.

Renee M. Clary is codirector and cofounder of the EarthScholars Research Group, currently located at Louisiana State University. She also teaches online geoscience courses for Northwestern State University of Louisiana. Dr. Clary does national and international archival and field-based research to improve geoscience learning in both formal and informal settings. She focuses on finding visualization- and technology-based solutions to critical problems that students and the public face in understanding geobiological concepts, principles, and theories. She enjoys probing the history of geology and specializes in the Golden Age of Geology. She is active in the Geological Society of America and the Geological Society of London's History of Geology Group.

Kathryn F. Cochran is professor of educational psychology in the School of Psychological Sciences at the University of Northern Colorado. She is coordinator of Graduate Programs in Educational Psychology and is on the advisory boards of the Mathematics and Science Teaching Institute and the Center for Collaborative Research in Education. Her interests include individual differences in learning in science, sociocultural views of learning and cognition, and philosophical perspectives on educational theory and research. She teaches Educational Psychology for Teachers; Learning, Cognition, and Instruction; and Research Issues in Cognition and Development. Her research interests include learning communities in online settings, pedagogical content knowledge development, and metaphor in teaching and learning.

Isobel R. Contento is the Mary Swartz Rose Professor of Nutrition and Education in the Department of Health and Behavior Studies, Teachers College, Columbia University, New York. She has served on advisory committees for the Centers for Disease Control and Prevention, the U.S. Department of Agriculture, and the Institute of Medicine. Her research interests are focused on factors influencing food choice and decision-making processes, particularly among children and adolescents; and the use of explorations about the impact of food and the food system on personal health and the environment as a basis of inquiry-based science education.

Marcia Dadds, registered dietitian, New York State certified dietitian nutritionist, and American Council on Exercise (ACE) certified personal fitness trainer, has been successfully educating and coaching individuals and groups to meet their nutrition and fitness goals for the past fifteen years. She holds an M.S. in nutrition education from Teachers College, Columbia University, and professional fitness certifications from Marymount Manhattan College and the American Council on Exercise (ACE).

Donna DeGennaro, a recent University of Pennsylvania graduate, is currently assistant professor of educational technology at Montclair State University. Donna's dissertation focused on how technology facilitates public-private school partnerships and empowers children across socioeconomic groups. Her research interests center on how both youth technology practices and interactions and cognitive, social, and cultural dimensions of learning inform innovative designs of learning environments. She is also interested in the relationship between aspects of educational organizations (leadership, teaching, learning, and professional development) and the adoption of technology-based innovations.

John Robert de Laeter is Emeritus Professor of Physics at Curtin University of Technology in Perth, Western Australia, where he was formerly vice president of research and development. He was originally a science teacher at the secondary level. His current research interests are in science education and in astrophysics and nuclear physics, and he has a long-standing interest in nuclear astrophysics and the origin of the chemical elements. Professor de Laeter is recognized as a science educator with a particular concern for ensuring that the public is informed of the role of science and technology in society. He was instrumental in establishing the Scitech Discovery Centre and the Einstein Cosmos Centre in Western Australia, both of which are successful science education centers for schoolchildren and the public.

Lisa A. Donnelly is a science education doctoral student in the Department of Curriculum and Instruction at Indiana University Bloomington. Lisa is a former high school biology teacher and presently teaches Elementary Science Methods for preservice science teachers. Her research interests are focused on the teaching and learning of biological evolution and teachers' use of state and national science standards as they make instructional decisions.

Danielle Dubno is a biology teacher at the Institute for Collaborative Education, a small, progressive, public secondary school in New York City. She recently received a master's degree in teaching biology from New York University. She was formerly research coordinator in a stem cell laboratory at Mt. Sinai School of Medicine and has conducted research on the marine life of the Turks and Caicos Islands. She is interested in learning more about how to use inquiry and long-term projects to teach science, and in finding effective ways to teach to diverse learners.

Garrett Albert Duncan is associate professor in the College of Arts and Sciences at Washington University in St. Louis. He holds appointments in education, African and African American studies, and American culture studies, and is director of doctoral studies in education. Formerly a middle and high school science teacher, Duncan is a recipient of the California Mathematics, Engineering, Science Achievement (MESA) Statewide Teacher of the Year Award; the Christa McAuliffe Fellowship Award, which supported the Pomona (CA) Saturday Science Academy; and a National Science Foundation Teacher Fellowship to pursue advanced studies in biology. His current research focuses broadly on matters of race and education in public schools.

Karen Elinich is director of Educational Technology Programs for the Franklin Institute Science Museum, Philadelphia. Her expertise is in the field of educational technology, specifically relating to the use of the Internet in support of science learning. Ms. Elinich joined the staff of the Franklin Institute in 1994 to participate in the leadership of a national project, known as the Science Learning Network, which explored how science museums could use the Internet to support teachers as they implemented science inquiry practice in their classrooms. Currently, she is developing the institute's educational presence for the K–12 Internet2 community.

Rowhea Elmesky is an assistant professor of science education at Washington University in St. Louis. In 2003, she received the American Educational Research Association (AERA) Division G, Social Context of Education Outstanding Dissertation Award. In 2004, Elmesky was the recipient of Paper Award V: Implications of Research for Educational Practice from the Association for Science Teacher Education (ASTE). Her main research interests lie in improving science teaching and learning opportunities for culturally marginalized and economically disadvantaged children. Her recent publications include a coedited book, *Improving Urban Science Education: New Roles for Teachers, Students and Researchers* (with Kenneth Tobin and Gail Seiler, 2005).

Melissa Fennemore is an inclusion and science teacher at Middletown High School in Delaware. Melissa began coteaching with student teachers from the

University of Delaware; and with Sue Gleason, she has pioneered coteaching as a model for inclusion classes.

First State Robotics, Inc., is a nonprofit organization based in Delaware and dedicated to exciting and inspiring students of all ages to pursue careers in science and technology. Adult volunteers were awarded the Governor's Award for Volunteerism in 2004 for their efforts. The organization runs a high school robotics team, which competes in the FIRST Robotics Competition; organizes robotics competitions in their local area; and holds the official First State Robotics Tournament, which features teams of students from elementary through high school who compete in FIRST LEGO League, Junior FIRST LEGO League, and Vex Robotics competitions.

Barry J. Fraser is professor and director of the Science and Mathematics Education Centre at Curtin University of Technology in Perth, Western Australia. This center has the world's largest doctoral program in science education. In 2003, he was the recipient of the Distinguished Contributions to Science Education through Research Award from National Association for Research in Science Teaching (NARST) in the United States. His research interests focus on learning environments and educational evaluation, and he is the current editor-in-chief of *Learning Environments Research: An International Journal.*

Pamela J. Garnett is the dean of curriculum at St Hilda's Anglican School in Perth, Western Australia, and teaches chemistry to final-year high school students. She is a winner of the prestigious Prime Minister's Award for Science Excellence (secondary teaching) and has achieved awards for excellence in science education research. Pam's interests are varied, encompassing research in chemistry misconceptions, exemplary teaching practice, science investigations, gender issues in science, the supply and quality of science teaching, and whole-school issues such as values education.

Cassondra Giombetti is a Ph.D. candidate in science education at the University of Pennsylvania's Graduate School of Education. Cassondra has experience teaching environmental science and experiential education in many settings, including informal, hands-on, and classroom environments. Her research interests include environmental education in urban settings, engaging student voice in curriculum development, and empowering students through science education.

Susan Gleason is a high school chemistry teacher and science department chair at Middletown High School in Delaware. Sue is active in statewide reform efforts in science education and is also involved with preservice science teacher education at the University of Delaware. With Melissa Fennemore, she has pioneered coteaching as a model for inclusion classes.

Brian Hand is professor of science education at the University of Iowa. He is coordinator of the Science Education Program, including the graduate and undergraduate programs. His research interests are in the area of science literacy, including the implementation of writing-to-learn strategies within science classrooms, impacts of student performance when using language as a learning tool, and exploring the necessary pedagogy for success.

William G. Holliday is professor at the University of Maryland, earlier at the University of Calgary. He served as executive secretary and later as president of the National Association for Research in Science Teaching. He has a B.S. and M.S. in biological sciences from Purdue University and a Ph.D. in science education from the University of Texas at Austin. His research interests include motivation-achievement, reading, studying, and textbook-based programs in science education, balancing implicit and explicit approaches to science teaching, and assessing influential studies and theories about learning. His practitioner efforts include publishing in NSTA's four regular periodicals.

Sophie Homan is currently employed by Rye Country Day School in Rye, New York, as an eighth-grade Introductory Physical Science teacher. In May 2005, Sophie completed her master of arts in science education at New York University's Steinhardt School of Education, where she graduated summa cum laude. Before attending NYU, Sophie worked at Rochester Institute of Technology for the Forest Fire Imaging Experimental System (FIRES) team in the Imaging Science Department while taking Ph.D.-level courses. Sophie graduated in 2003 from the State University of New York at Geneseo with a bachelor of arts in physics.

Susan A. Kirch is an assistant professor of science education in the Elementary and Early Childhood Department at Queens College of the City University of New York. Her research interests include developing a comprehensive account of children's abilities to orchestrate and participate in the process of scientific inquiry. She also studies the uses of coteaching as a way of providing professional development in learning to teach using scientific inquiry. She teaches Research Issues in Math, Science, and Technology Education; Science for the Elementary School Teacher; Life Science for Elementary School Teachers; and Environmental Science for Elementary School Teachers. She also coteaches courses in biomedical research in the Biology Department.

Christine (Kit) Klein conducts research and evaluation in informal science-learning settings as an independent consultant. Prior to this, Kit worked in various formal and informal education settings, most recently serving as the first principal investigator of the St. Louis Center for Inquiry in Science

Teaching and Learning funded by the National Science Foundation in 2001. Her interests have always focused on connections between informal and formal science learning. In 1996, she cofounded the Informal Learning Environments Research Special Interest Group of the American Educational Research Association to encourage and support other researchers in this work.

Steve Kluge has been teaching earth science and geology at Fox Lane High School for twenty-seven years. His classes and labs are driven by current events, and are designed to lead his students to their own discovery of facts and principles that explain the world around them. He considers field experience essential, and leads several student trips each year. In recent years, Steve has been leading workshops and seminars for earth science teachers at State University of New York, Purchase College, and he is involved in several state and national efforts to improve geoscience education at the high school level. Steve has twice been recognized as Outstanding Earth Science Teacher by the National Association of Geoscience Teachers. He measures his success in the classroom by the significant number of his former students who have become his friends and colleagues as geoscientists and geoscience educators.

Pamela A. Koch is a Research Associate at Teachers College Columbia University. She has worked as a registered dietician, teacher, and nutrition/ science education researcher. Her research specializes in designing inquiry-based curriculum focused on science and nutritional literacies and its impact on student health and development. Her work has been published in the *International Journal of Science Education,* and *Science and Children,* among other places. She is the project manager of Life, Linking Food and the Environment, a nonprofit teaching and research center at Teachers College.

Richard H. Kozoll is an assistant professor of science education at the School of Education of DePaul University. His research interests include students' identity constructs as a means to understand engagement with and participation in science and inform inclusive science teaching practices. An additional line of research includes the use of narrative in science education research.

Joseph Krajcik, a professor of science education at the University of Michigan, focuses his research on designing science classrooms so that learners engage in finding solutions to meaningful, real-world questions through inquiry. He has authored and coauthored over 100 articles or chapters. He makes frequent presentations that focus on his research as well as those that translate research findings into classroom practice. His colleagues have recognized his leadership abilities by selecting him president of the National Association for Research on Science Teaching. In 2001, the American Association for the Advancement of

Science recognized and inducted him as a fellow. Prior to obtaining his Ph.D. from the University of Iowa, he taught high school chemistry.

Sarah-Kate LaVan is an assistant professor of science education at Temple University in Philadelphia, Pennsylvania. Her research interests involve examining the development of classroom communities and structures that foster equity, agency, and collective responsibility. In particular, she focus on three distinctive areas within the field of science education: (1) the examination of resources (human, material, and symbolic) as structures; (2) the use of cogenerative dialogue, video analysis, and social theory as a means to expand the agency of teachers and students; and (3) microlevel understandings of the development of culture and science fluency.

Kimberly Lebak is assistant professor of education at the Richard Stockton College of New Jersey. Her research interests include connecting the teaching and learning of science in classroom settings to informal learning settings and the use of cogenerative dialogues as a structure to maximize potential learning opportunities for teachers and students.

Miyoun Lim is a Ph.D. candidate in science education at Teachers College, Columbia University. Her dissertation explores urban children's sense of place and its implications for urban science education. She is currently working on developing an online teacher environmental education program using urban (New York City) ecosystems. Her research interests include teaching and learning science in urban schools, especially those with immigrant minority students living in high-poverty neighborhoods, and urban environmental education—informed by critical pedagogy of place—for students and teachers.

Jessica J. Luke is a senior researcher at the Institute for Learning Innovation, a not-for-profit learning research and development organization in Annapolis, Maryland. Her research is focused on how and what people learn outside of school, particularly related to youth development and school-family-community interactions. She has published numerous articles and book chapters in these areas. As an adjunct faculty member at George Washington University, Washington, D.C., she teaches a graduate course in museum evaluation.

Rachel Mamlok-Naaman is head of the National Center for Chemistry Teachers at the Weizmann Institute of Science, and a senior staff scientist in the Chemistry Group. She is engaged in development, implementation, and evaluation of new curricular materials, and research on students' perceptions of chemistry concepts. Her publications are in the areas of scientific literacy, curriculum development, teachers' professional development, and cognitive aspects of students' learning.

Sonya N. Martin is an assistant professor of science education at Queens College in Flushing, New York. Martin taught science at both the elementary and secondary levels in the Philadelphia Public School District for five years, during which time she earned two graduate degrees in elementary education and chemistry and education from the University of Pennsylvania. At Queens College, Martin works with preservice and in-service teachers who are learning to teach science in the elementary school. Her research interests are focused on the cultural and social dimensions of teaching and learning science in the elementary classroom and on developing teachers as researchers of their own teaching practices.

Dale McCreedy is the director of Gender and Family Learning Programs at the Franklin Institute Science Museum. Over the last seventeen years, she has implemented a range of national programs and collaborations with community-based organizations that targeted girls, families, and underserved communities. Recent research efforts have focused on girls and women's trajectories in science careers and hobbies as a result of out-of-school learning experiences. As an advocate for girls and women's science learning, Dale participates on numerous advisory boards, and was the 2002 winner of the Maria Mitchell Award for Women in Science.

Judith A. McGonigal is presently an eighth-grade science teacher in the Haddonfield Public School District in New Jersey, where she has previously taught kindergarten, first, and fifth grades. She also is an adjunct instructor in the professional development program at the University of Pennsylvania Graduate School of Education and a lecturer in the teach preparation program at Rutgers University, Camden. As a teacher-researcher and reflective practitioner, her focus has been on documenting how communities function as colearners, coteachers, and co-researchers to construct their own best practices for science and literacy learning.

Catherine Milne is an assistant professor in science education at New York University. She is currently involved in a large research study, titled "Molecules and Minds," funded by the U.S. Department of Education to examine the development and use of simulations and animations of chemical phenomena and explanations in schools. Her other research interests include urban science education, the nature of coteaching, self-evaluation strategies, and the nature of science in science education.

Ji-Myung Nam came from South Korea for her graduate study at New York University. She is majoring in science education for teaching middle school and high school students. Her specialty is biology. She is interested in multicultural science education and making effective diagrams and visual materials for understanding contexts.

Sharon Nichols is an associate professor of science education in the College of Education at the University of Alabama. Over the twelve years of her career in science teacher education, she has developed coteaching relationships with "teacher-friends" to maintain a grounded sense of what it means to teach and learn science in today's classrooms. Since moving to Alabama in 2002, she has focused on understanding science education as a cultural practice involving African American teachers and students. Her insights regarding cultural diversity and science education have been extended through work with teachers in Australia, Philippines, Colombia, and Peru. She has drawn on sociocultural and feminist perspectives to explore teachers' science learning and teaching practices re-presented through narrative and visual ethnographies.

Loaiza Ortiz completed her Ph.D. in science education at Teachers College, Columbia University, and wrote her dissertation on holistic understandings of how urban youth appropriate science practices. She is currently working with science and nutrition education colleagues on a curriculum that integrates science, nutrition, and environmental education. Her professional interests include sociocultural understandings of teaching and learning science in urban contexts, and urban environmental and ecology education, particularly related to food and food systems.

Margery D. Osborne is an associate professor in the College of Education at the University of Illinois at Urbana-Champaign. She teaches early childhood and elementary science education courses. Her research interests include exploring the dynamic and complex nature of teacher knowledge and how they are located within the intersections constructed between ideas of reflective practice and research on critical and feminist pedagogy.

Tracey Otieno became involved in conducting classroom research through her association with the Discovering Urban Science research group while teaching high school chemistry and physics in the Philadelphia Public School District. After completing a master's degree in chemistry education at the University of Pennsylvania, Tracey continued working for the program as an internal evaluator.

Lilian Pozzer-Ardenghi is a Ph.D. student in curriculum studies at the Faculty of Education, University of Victoria, British Columbia. She graduated in biological sciences at the Universidade Federal de Santa Maria, RS, Brazil, then pursued a master of arts in education at the University of Victoria, investigating the use of photographs in high school science textbooks and lectures, and students' interpretation of photographs present in high school biology textbooks. Currently, her research focuses on nonverbal communicational aspects of science teaching, particularly gestures.

Scott Ritchie is a Ph.D. student in elementary education at the University of Georgia and an instructional coach for the Clarke County School District in Athens, Georgia. In 2003, Ritchie served as a consultant to the National Board of Professional Teaching Standards, and in 2001, Ritchie was recognized by CNN as one of "America's Heroes" for his teaching practice. His research interests focus on critical literacy and multicultural education.

Stephen M. Ritchie is an associate professor in science education at Queensland University of Technology, Brisbane, Australia. Steve's research has focused mostly on classroom issues that relate to teaching and learning science. His recent research projects are concerned with leadership dynamics within high school science departments, teacher change, science teacher education, research collaboration, student science learning through the co-creation of eco-mysteries, and student research projects on sustainable development.

David Robison has been teaching Regents Earth Science at Wilson High School for ten years. Prior to teaching at Wilson, he served as a Peace Corps Volunteer in Honduras, and upon his return taught science and Spanish at an alternative high school in New York City. David's classes are driven by current events, and are designed to help his students observe their world more carefully in order to become more critical thinkers. An award-winning teacher and leader in the earth science community, David has expanded his influence through his Science Teachers Association of New York State Share-a-Thon website (www.regentsearthscience.com/webshare) and by conducting many Science Teachers Association of New York State leadership workshops. David hopes to one day have half as much of the impact on his students and colleagues that his friend and coauthor Steve Kluge has made in the field of earth science.

Alberto J. Rodriguez is associate professor of science education in the Department of Policy Studies in Language and Cross Cultural Education, and the codirector of the Center for Equity and Biliteracy Education Research at San Diego State University. Dr. Rodriguez teaches bilingual science methods courses in the bilingual teacher certification program, and he also teaches graduate courses in the master and doctoral programs. His research interests are bilingualism, multicultural education, and sociotransformative constructivism.

Wolff-Michael Roth is Lansdowne Professor of Applied Cognitive Science at the University of Victoria, British Columbia. After teaching middle and high school science, computer science, and mathematics for over a decade, he began his university career as a statistics professor prior to taking his current position. He now conducts research into knowing and learning mathematics and science from kindergarten to professional practice, which is published in linguistics, sociology, and education journals. His recent publications include

Toward an Anthropology of Graphing (2003); *Talking Science: Language and Learning in Science Classrooms* (2005); and, with A. C. Barton, *Rethinking Scientific Literacy* (2004).

Katy Roussos is a special education teacher at Elizabeth Haddon Elementary School in Haddonfield, New Jersey. She specializes in the area of inclusion, developing programs that meet the needs of special education students in regular elementary school classrooms while creatively collaborating with classroom teachers. The success and enthusiasm of her students once reintegrated into the typical classroom setting are testaments to her unique approach to problem solving and differentiated teaching.

Paul H. Ruscher is associate professor of meteorology at Florida State University, where he also serves as associate chair and director of Outreach and Undergraduate Programs. His research emphasizes coastal meteorology, and he has been an active developer of innovative teacher professional development programs in the geosciences. He is a former principal investigator for the GLOBE program, and he won the Society for Information Technology and Teacher Education Award for Excellence in Science Education in 2001.

Kathryn Scantlebury is an associate professor of chemistry and secondary science education coordinator at the University of Delaware. Her research interests focus on gender equity in science education, the use of coteaching in preparing science teachers, and the effectiveness of professional development programs.

Gale Seiler is an assistant professor at the University of Maryland Baltimore County, where she teaches science education and multicultural education courses. Her research in Baltimore city schools addresses culturally specific way in which inner-city, African American students participate in school science. She is also interested in preparing teachers to teach students in urban schools. Her doctoral dissertation received the Distinguished Dissertation Award from the Association of Teacher Educators. Before completing her Ph.D., Gale was a high school science teacher for sixteen years.

Harry L. Shipman is the Annie Jump Cannon Chair of Physics and Astronomy at the University of Delaware. He has two distinct research agendas in astronomy and in science education. In astronomy, he has studied white dwarf stars for several decades, and most recently has discovered unexpected clouds of hot gas around both white dwarf stars and the very tiny "brown dwarfs," objects that are too small to be stars and too big to be planets. In science education, his interests are the application of inquiry methods to large classes, teaching the nature and process of science, and multicultural science education. He teaches a wide variety of students with particular emphasis on preservice and in-service teachers. He is also a competitive figure skater. In

2002, the National Science Foundation recognized him as one of six recipients of the Distinguished Teaching Scholar award.

Ruby Simon has taught sixth-grade science for twenty-two years in Tuscaloosa, Alabama. She is currently teaching science to African American students of multiability levels who are predominately low socioeconomically and disadvantaged. Over the past several years, she has focused on developing leadership roles among students, teachers, and community members as a means to enhance teaching and learning in classrooms. She often spends nights and weekends recruiting and working with students and volunteers from local churches and organizations to sustain community-based support for the school. Currently, she is participating in a research study group supported by a National Education Association grant to develop approaches to teaching culturally relevant science and literacy.

Marsha Smith is currently a biology teacher at William H. Maxwell Career and Technical Education High School. She received a master's degree in science education from New York University. She recently placed first in the Region 5 Technology Festival for designing a web page with her students. This team also made it to the Semifinals for the Thinkquest 2005 NYC competition. Marsha is looking forward to learning new and innovative ways to enhance learning while incorporating technology into her classroom.

Michael J. Smith began his career twenty years ago as a geologist, conducting glacial marine studies in Antarctica, exploring for oil and natural gas in Africa, and doing groundwater investigations of hazardous wastes sites throughout the eastern United States. In 1988, he became an earth science teacher and went on to earn his doctorate in science education. In 1991, he was selected as the Outstanding Earth Science Teacher of Pennsylvania by the National Association of Earth Science Teachers. From 1998 to 2004, he was director of education at the American Geological Institute in Alexandria, Virginia, where he led the EarthComm, Investigating Earth Systems, and Constructing Understandings of Earth Systems (CUES) curriculum programs. Mike teaches seventh-grade earth science at Wilmington Friends, and teaches the middle school science methods course within the University of Pennsylvania's Masters in Integrated Science Education Program. He is the editor of *The Earth Scientist*, the quarterly journal of the National Earth Science Teachers Association.

John B. Southard has been a professor of geology at the Massachusetts Institute of Technology since the late 1960s. His research specialty is the physics of sediment transport and the interpretation of ancient sedimentary environments. His work has included both field and laboratory studies of sediments, sedimentary processes, and sedimentary rocks. He has taught courses in introductory geology and sedimentary geology, both at MIT and in Harvard's evening adult education program. For many years, he has been technical editor of the *Journal of Sedimentary Research*, the flagship journal of the Society for Sedimentary Geology (SEPM). He was senior writer for EarthComm and Investigating Earth Systems, curriculum programs produced the American Geological Institute for use in secondary schools.

John R. Staver is professor of education (science) and director of the Center for Science Education at Kansas State University. He was elected a fellow in the American Association for the Advancement of Science in 1994 for his work on behalf of a national reform agenda in science education. His research focuses on constructivist epistemology and its implications for improving science teaching and learning. He is also examining the interface between science and religion within a constructivist perspective, with a focus on the nature of each discipline and conflict between them over evolution.

Deborah J. Tippins is a professor of science education in the Department of Mathematics and Science Education at the University of Georgia. She holds a joint appointment in the Department of Elementary and Social Studies Education and serves as program head for Elementary Science. A recent Fulbright scholar in the Philippines, she developed a longitudinal, collaborative research program entitled "Transforming Science Education for the 21st Century." Her research interests involve anthropological approaches to science education with a focus on community-centered and culturally relevant science teaching and learning. Additionally, her research interests include science teacher learning through case-based pedagogy, and she has co-authored three casebooks for elementary and secondary science teachers.

Barbara Tobin is an education consultant with specialization in children's literature, drawing on a wealth of classroom experience at all levels, across four continents, and over four decades. She has worked extensively with preservice and in-service teachers, mostly in literacy and children's literature. She is the recipient of the 2003 Excellence in Teaching Award from the Graduate School of Education at the University of Pennsylvania. She currently serves on the review board of the International Reading Association's journal, *The Reading Teacher*.

David F. Treagust is professor of science education at Curtin University of Technology in Perth, Western Australia, where he teaches courses in campus-based and international programs related to teaching and learning science. His research interests include understanding students' ideas about science concepts and how these ideas relate to conceptual change, the design of curricula, and teachers' classroom practices. He is a member of the Australian National Advisory Committee for Program for International Student Assessment, was president of the National Association for Research in Science Teaching (1999–2001), and is currently managing director of the Australasian Science Education Research Association.

James Truby is a middle school science teacher at Rock Lake Middle School in Florida. An award-winning teacher, Jim is known for his inspirational teaching of cutting-edge science activities that actively involve his students with inquiry, leading scientists around the United States, and the community. A testimony to his talent and enthusiasm for science teaching is that many of his students go on to pursue science at higher levels and seek employment in science-related fields.

Purvi Vora is a Ph.D. candidate in the science education program at Teachers College, Columbia University. Her research focuses on preparing secondary science teachers to teach for social justice in urban schools. She has a Master of Science in Microbiology and Molecular Genetics from Rutgers University. She is the co-creator of an online, multimedia case-based environment called Project Yuva that presents teachers with three cases around student agency, ownership, and funds of knowledge of urban, middle school youth in New York City. She is part of the Social Justice Collaborative, a research collaborative established to encourage junior scholars and mentors to pursue, share, and present research related to urban science education.

James H. Wandersee is the William LeBlanc Alumni Association Professor of Biology Education in the Department of Curriculum and Instruction at Louisiana State University. He is an elected fellow of the American Association for the Advancement of Science (AAAS) in the Biological Sciences section and is a regular presenter at scientific meetings as well as science education meetings worldwide. His research focuses on finding and testing innovative visual and historical approaches for improving public understanding of science, particularly in the fields of plant biology and geobiology. He is founder of the 15° Laboratory for visual cognition studies and cofounder of the EarthScholars geobiology research group.

Beth Wassell is an assistant professor in the Department of Secondary Education at Rowan University in Glassboro, New Jersey. Currently, her courses focus on general teacher education, foreign-language teaching methods, and linguistic and cultural diversity. Her research interests range from teacher identity and beginning teachers' experiences in urban schools to the needs of English language learners in high school contexts.

Randy K. Yerrick is a professor of teacher education and research fellow at the Center for Research in Mathematics and Science Education at San Diego State University. In 2000, Yerrick was recognized as an Apple Distinguished Educator for his work in shaping the use of contemporary instructional technology in science classrooms. His research interests are focused on the teaching and learning of science among diverse student populations, with a focus primarily on equity issues for African American and Latino/Latina students.